QUALITY ASSURANCE

OF

FOOD

Ingredients, Processing and Distribution

QUALITY ASSURANCE
OF
FOOD
Ingredients, Processing
and Distribution

by

John E. Stauffer, Ph.D.
STAUFFER TECHNOLOGY
GREENWICH, CONNECTICUT

FOOD & NUTRITION PRESS, INC.
WESTPORT, CONNECTICUT 06880 USA

FOOD & NUTRITION PRESS, INC.
Westport, Connecticut 06880 USA

Library of Congress Catalog Card Number: 88-080008
ISBN: 0-917678-23-0

Printed in the United States of America

DEDICATION

In Memory

of

HANS STAUFFER

whose genius, drive and vision

shall always be an inspiration

CONTENTS

ACKNOWLEDGMENTS

A word of appreciation must be expressed to my former associates at Stauffer Chemical Company. Without their support over the years, I never would have been inspired to write this book. Six individuals deserve special mention because of their knowledge and experience: Robert S. Bryant, James T. Elfstrum, Joseph V. Feminella, Conrad S. Kent, Harold B. Reisman, and Lorna C. Staples. Their forbearance, good nature and willingness to tolerate a chemical engineer in their midst will always be remembered.

My indebtedness to my publisher, John J. O'Neil, cannot go unmentioned. He provided helpful suggestions on specific sections and on the general layout of the book.

Finally, I wish to acknowledge the encouragement, assistance, and valuable critique received from my wife, Valerie.

PREFACE

No food processing organization can function successfully without quality assurance. It is the essential discipline needed to guarantee a safe, wholesome, and functional product. From the first conception of a new product, quality assurance must be made an integral part of ingredient testing, processing, and distribution. The customer must be satisfied that every precaution has been taken to assure product excellence.

From the outset, a distinction needs to be drawn between Quality Assurance and Quality Control. Unfortunately, because these two terms have been used interchangeably the difference between them has become blurred. Quality Assurance can be defined as a strategic management function which establishes policies related to quality, adopts programs to meet the established goals, and provides confidence that these measures are being effectively applied. Quality Control is a tactical function which carries out those programs identified by Quality Assurance to be necessary for the attainment of the quality goals.

The aim of this book is to describe the scope, applications, and benefits of quality assurance. For the sake of clarity, the subject is divided into four parts: Product Planning, Manufacturing, Customer Service, and Distribution. Within each section, selected topics of vital importance to quality assurance are discussed; these chapters include Good Manufacturing Practice, Hazard Analysis and Critical Control Points, Kosher Certification, Packaging, Labeling, and Product Recall. The practices outlined in this book have been tried and proven in industry, and they represent the best tools available for achieving the highest standards.

When properly conceived and executed, quality assurance is a strong asset to an enterprise. Rather than throwing up roadblocks in the way of progress, this discipline is designed to help a manufacturer expedite the production of quality products. By specifying those procedures which must be routinely followed, an effective quality assurance program helps to avoid much unnecessary, redundant, and tedious activity. Moreover, time and again such programs have been demonstrated to be financially sound. Doing the job right the first time is good business. Any program which helps to reduce losses and promotes efficiency can only have a positive effect on profits.

In advance, a few caveats are in order. The reader looking for an exhaustive discussion of a given topic may be disappointed. That person is referred to the many excellent monographs in print. Other readers may become impatient with historical and anecdotal material included throughout the text. Entertaining as these accounts may or may not seem, their only purpose is to show the relevancy of basic concepts. And finally, those readers, who have an aversion to legal phraseology, are reminded that existing government regulations, which codify many of the principles of quality assurance, must be accepted as the final authority.

This book is written for and dedicated to all quality assurance personnel whose responsibility it is to maintain a safe food supply. It is directed at those persons charged with making the day-to-day decisions, whether working in industry or government. This volume also has special significance for those selfless instructors in academia whose role is to train the future managers and inspectors. The success of this book will ultimately be measured by how well it serves the requirements of these individuals for whom it was intended.

JOHN E. STAUFFER

FOREWORD

A country's food industry, at whatever level of complexity it has achieved, provides a basic product. Whether grown in one's backyard, obtained by barter from a neighbor's farm or purchased in a corner delicatessen or shopping mall supermarket, we cannot do without it. Even when we seek the convenience of another's final preparation in a fast food outlet or enjoy gastronomic delights in a *haute cuisine* restaurant, we are satisfying a most fundamental and indispensable need. This may, upon reflection, seem obvious, but in our modern technological society, fulfilling our requirement for sustenance is neither simple nor easy.

Few, if any, can live exclusively on the fruits of their own labor, for our farms and ranches today are too specialized to provide all the ingredients of a "balanced diet." Thus over the years we have become almost entirely dependent upon others to supply our food, giving birth to today's multibillion dollar food industry. We rely on its rapid and often highly complex means of production, processing, marketing and distribution to provide us with a daily array of products that are appealing, affordable and wholesome. Underlying these desirable and important attributes there is, we assume, quality.

There is no single definition which immediately flows from the concept of food quality. In various contexts it can mean purity, safety, value, nutrition, consistency, or honesty, e.g., in labeling. This list is certainly not exhaustive. But whatever the circumstance, we presume there are standards which, when met by a particular product, define its quality. There was a time when standards were obvious and immediately verifiable. We harvested our own wheat certain it was not contaminated or admixed with other grains. A neighbor's steer gave us what we knew was only fresh sirloin since we saw it dressed. Now the situation is more complicated, for food manufacturers rely on multi-source ingredients often from distant points. Processing utilizes sophisticated methodology and distribution is rapid through complicated marketing channels. Consumer needs and expectations have given rise to the use of added flavorings, colors, and preservatives. The economics of farming have led to expanded use of fertilizers and pesticides. Nutritional concerns raise questions about macroingredients like protein and fat, and microingredients like vitamins and sodium.

These and related aspects of food quality demand standards, some of which are determined within the food industry and others by the many government agencies which function with legislative mandates. These latter are, in fact, congressional interpretations of consumer needs and expectations. Thus from various sources there are hosts of prescriptive and proscriptive standards to which those in the many food companies, large and small, must adhere.

How they do it is quality assurance. It is to this monumental subject, its history, strategies, methodologies and pitfalls that this book is addressed. Although comprehensive in scope, this volume has a focus which is sharp and unwavering. Logically organized it begins with product planning and proceeds through manufacturing and customer service to distribution. Critical areas such as good manufacturing practices (GMP's), packaging and labeling, and product recall are explored with illustrative examples and, where appropriate with tables, diagrams and charts. Clearly written and well-documented, this is at once a textbook, manual, and reference work. Every reader will not agree with all the author's conclusions and suggestions, but were such not the case, a book on this topic would be banal and superficial. This volume, in short, is a guide which is both stimulating and challenging. Its publication is an important contribution to the literature of the quality assurance of our food supply.

ARTHUR HULL HAYES, JR., M.D.
Formerly Commissioner,
U.S. Food and Drug Administration

PART I

Product Planning

CHAPTER ONE
ORGANIZATION AND MANAGEMENT

In the absence of clearly defined responsibilities within a corporation, quality assurance becomes everybody's business and nobody's business. Quality assurance is the concern of everyone, from the newly hired shift operator to the Chairman of the Board, but without an effective corporate organization these individuals, by default, all too often will fail to perform the tasks required of them. Recognizing the need to hold managers answerable, the Food and Drug Administration (FDA) stipulates in the *Code of Federal Regulations*, 110.10(d) that:

> Responsibility for assuring compliance by all personnel with all requirements
> of this Part 110 shall be clearly assigned to competent supervisory personnel.

Individual responsibilities must be delineated in formalized procedures that cover the different functions of quality assurance. Until decisive measures are taken to enforce specific objectives, any quality assurance program will remain a well-meaning but ineffective body of rhetoric and platitudes. Procedures should be concise, lucid statements that pinpoint job assignments and provide means for communication within the organization. In the preparation of such directives, a realistic appraisal should be made of the needs of the business and the resources and manpower available for meeting these demands. A quality assurance manual comprised of these procedures should be distributed to all supervisory personnel.

Organization requires careful thought and deliberation; this necessity will be made clear in the following sections of this chapter. Left to the whims of management or the vicissitudes of mergers, acquisitions, and divestitures, corporate structure soon becomes disordered. Haphazard planning can only result in people working at cross purposes. No other assignment in quality assurance is more important than developing a strong organization in terms of both ability and mission.

Corporations are not the only principals concerned with organization and management. Industry's counterparts, the government regulatory agencies, manage staffs that outnumber the employees of most food processors. These civil servants are involved with the same food-related issues, albeit from a different point of reference. What the government regulators say and do has as great a bearing on product planning as any decisions made by industry personnel.

PARK DECISION

Two decisions by the United States Supreme Court have far-reaching consequences for food processors and the conduct of their affairs. These cases unequivocally hold the chief executive officer responsible for the actions of his subordinates. This tenet is unabridged by any consideration of company size, good intentions, lack of knowledge, or other mitigating circumstances.[1]

In the Dotterweich case that goes back to 1943, the doctrine of "strict" liability, sometimes referred to as "absolute" or "vicarious" liability, was enunciated. In this landmark case the court held that the president of a small drug repackaging establishment could be found guilty of a violation of the Federal Food, Drug and Cosmetic Act (FD&C Act) even though he personally may not have been involved with the given transgression. The decision took cognizance of an unusual feature of the FD&C Act, namely, its failure to specify any element of intent, knowledge, or willfulness as a prerequisite to criminal prosecution. Thus, a defendant is barred from relying on any assertion that he was ignorant of an offense.[2]

The strict liability doctrine was upheld and expanded in a subsequent Supreme Court action. In 1975, the Park decision not only reaffirmed that lack of knowledge is an invalid defense, but it went on to state that persons in high corporate positions of authority can be held accountable for failing to take preventive measures.[3] The court found that John R. Park, President of Acme Markets, Inc., which operated more than eight hundred food stores, was guilty when his supermarket chain was condemned for rodent infestation in one of its warehouses. Chief Justice Warren E. Burger in writing for the majority maintained that "the requirements of foresight and vigilance" demanded of chief executives must be upheld.[4]

Commenting on the strict liability doctrine, the Department of Health, Education and Welfare (HEW), testified in a Senate hearing in 1976 as follows:

> Since the civil remedies available to FDA . . . are essentially retrospective in effect, food processors can, and too many do, simply sit back and wait for FDA to act. It is far cheaper to risk the loss of a few hundred or thousand dollars as a result of an occasional seizure or injunction than to regularly allocate many thousands of dollars necessary to maintain facilities in sanitary condition. The primary impetus to self-regulation is the fear that criminal penalties may be imposed for failure to take every precaution to ensure that violations — and their potentially harmful consequences — will not occur.[5]

The debate goes on whether it is fair to hold senior officers in major food corporations responsible for the errors of their underlings, but so far the court decisions stand. Congress has held numerous hearings on new food legislation during which this issue has continued to be aired. Some people feel that constitutional rights are being encroached by adhering to such a strict standard. The prevailing opinion, however, asserts that in the sensitive area of food safety, the public's welfare takes precedence over consideration of individual claims.

ORGANIZATION CHART

The court finding in the Park case, specifically, that a CEO can be held accountable for the actions of his subordinates, is very significant when addressing the question of organization. A prudent chief executive is compelled to establish clear channels of communication between the persons making the decisions at the plant level and those executives responsible for setting policy. Senior management should have access to operational data, and line supervisors should be able to report developments as they occur. Impediments to the exchange of information, whether built into the organization or unknowingly present, can only lead to misfortune.[6] A simplified organization chart is shown in Figure 1.1. It includes the primary functions of marketing, manufacturing, and quality assurance, but it does not attempt to designate all supporting groups which may vary from one company to another. While some variations from the basic structure are possible, several key features should be noted. The senior person responsible for quality

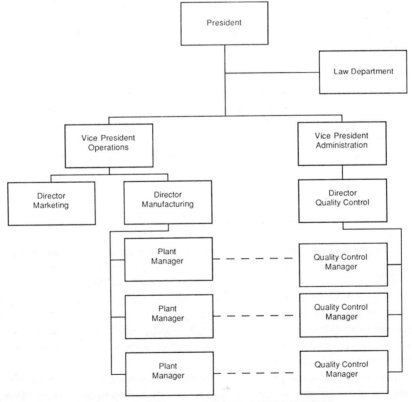

FIGURE 1.1. ORGANIZATION CHART

assurance reports directly to top management and has equal status with the heads of marketing and manufacturing.[7] The plant quality control managers each report to the Director of Quality Assurance although dotted lines on the organizational chart indicate close functional ties with the plant managers.[8]

In the quality assurance profession, strong feelings have been expressed about the need to have the plant laboratories report directly to Quality Assurance rather than to Production. If quality control is allowed to slip into the manufacturing organization, there may arise a tendency to overlook danger signals and to delay corrective actions.[9] For the quality control manager to discharge his duties effectively, he must be given wide latitude without fear of reprisals. When the quality control manager is exposed to subtle pressures or is subject to undue influence in hiring, firing, and salary administration, he loses an important sense of perspective.[10]

The cornerstones of any quality assurance program are the food scientists in the plant control laboratories. Upon their shoulders rest much of the burden to maintain a steady output of quality products. In keeping with the principle of decentralization, as much responsibility as possible should be delegated to the laboratories. Not only are these groups closer to the problems that beset quality, but unfailingly they are far more efficient than corporate staffs in handling the deluge of requests from customers.[11]

The Director of Quality Assurance is charged with exercising tight control over all aspects of quality. His accepted duties include the approval of all labels, packaging, product specifications, special releases of product, and data sheets. The organization depends more on the Director of Quality Assurance than any other person for implementing the necessary programs and procedures. The CEO can sleep better at night knowing that this critical job is in the hands of a competent professional.

RECALL ORGANIZATION

The events leading up to a product recall can engulf an organization with bewildering and traumatic swiftness. The company's response to such a challenge is crucial in order to avert confusion, indecision, and errors in judgement. How well management reacts in a recall situation depends in large measure on whether the company has taken certain preparatory steps in advance. Of foremost importance is the establishment of a standby recall organization. As soon as a crisis strikes, this organization is ready to respond immediately to the emergency and thereby head off potential disaster.[12,13]

While the mechanics of product recall are reviewed in detail in Chapter Thirteen, the organization used to implement a recall is described in this section. This discussion highlights the differences between a recall organization and the normal business structure which it supplants in times of such upsets. Whereas the company's usual way of conducting its affairs is suitable for making everyday decisions, that structure is inadequate to cope with a crisis. Responses are too slow and deliberate to meet effectively the developments which unfold in a matter of hours instead of days or weeks.

Central to any recall organization is the Recall Coordinator as shown in Figure 1.2. This position is filled by one person, not by a committee. The person usually selected for this assignment is the Director of Quality Assurance because of his familiarity with all concerns relating to quality. For the duration of the recall he assumes the powers of a czar; he is entrusted to make the necessary snap decisions, however controversial they may seem at the time. He is given access to the President's Office, bypassing the normal lines of communication.[14]

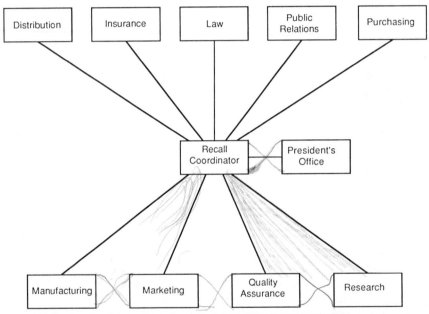

FIGURE 1.2. RECALL ORGANIZATION

Notwithstanding the authority vested in the Recall Coordinator, he does not act alone. Rather, he orchestrates a complex assembly of individuals consisting of line managers, administrative personnel, and representatives from staff departments. When called upon, all of these persons cooperate by providing information, giving assistance, and offering advice to the Recall Coordinator. Directly or through spokesmen, he coordinates all contacts with persons outside of the company whether they are customers, regulatory officials, or the press. To a significant degree the success or failure of a recall rests in the hands of the Recall Coordinator. The following account illustrates better than any exposition the lessons to be learned from an actual recall.

A Recall by Hygrade Food Products

When Hygrade Food Products Corporation of Southfield, Michigan, received a report on the last Wednesday in October, 1982, that customers in the Detroit area were finding razor blades in its Ball Park hot dogs, the meat packer desperately began recalling millions of frankfurters. In vain the company searched for some demented criminal either within its ranks or outside the organization. As the hours passed into days and the mystery remained unsolved, officers of the company began to fear for Hygrade's very existence. Alarmist reports about the incident were circulated in the newspapers and on TV. The Department of Agriculture (USDA), which was overseeing the massive recall, failed to uncover any clues. Finally, after days of frantic efforts, the case was resolved: the reports were all hoaxes planted by either pranksters, extortionists, or sociopaths.

From the beginning of the recall, Hygrade's Operations Vice President assumed responsibility for coordinating the investigation and acting as spokesman for the company. Key strategy decisions were made on the spot by him in consultation with the President. One important decision was regrettable, namely, to issue a press release supporting what later turned out to be a false lead. But the other actions which were taken effectively dispelled the rumors and helped to calm the hysteria. By its decisive and forthright stance, the company, in the end, more than regained its lost business and restored customer confidence in its products. This frightening occurrence vividly dramatizes the unexpected twists of events during a product recall, and it illustrates how a company like Hygrade, when properly prepared, can successfully confront the challenge.[15]

COMMITTEES AND TASK FORCES

Committees are a way of corporate life. Even if an agreement could be reached to banish them, there is some doubt whether this could be achieved. Instead of debating their merits, the more rewarding approach is to examine the special situations where committees are useful in quality assurance.

Because many disciplines are involved in the management of quality assurance, individuals who are expert on all subjects are difficult to find. Therefore a manager must rely on the input from many sources in order to assess his options adequately. The formation of standing committees whose members represent different backgrounds provides an effective means of bringing these skills to bear on particular problems. Committees, for example, can be established for reviewing government regulations, packaging, or labeling.[16]

While committees are well suited for communication, they have their shortcomings in decision making. Unless the chairman is given the necessary authority or by his station wields sufficient influence, he will be frustrated in attempting to reach a consensus on any particular issue. Because the forces within a committee are diffuse, they are not easily mobilized in support of a given course of action.

At the onset of problems which refuse to go away, management's frequent reaction is to form a committee. Such an ad hoc committee often is referred to as a task force, perhaps because these words have a better connotation. Charged with finding a solution, this body's ultimate aim is to work itself out of a job. The approach of forming such task forces is not wholly irresponsive. Many witnesses give testimony to its success, and others are quick to reason that hiring a consultant is a worse alternative.[17] In any event, the establishment of a task force focuses attention on the problem, allocates needed resources, and sets a timetable for reaching goals. If given a free hand, more often than not the task force will come close to expectations.

ENVIRONMENTAL AUDIT

Borrowing a concept from finance, manufacturers are beginning to establish environmental audit groups. These persons are charged with overseeing compliance in such areas as waste disposal, pollution control, toxic substances regulation, and occupational safety and health. By analogy this activity has been extended to include the surveillance of food quality assurance programs.[18]

There is no question about the need for a watchdog function to look over the shoulders of operating personnel. This double check assures management and other interested parties, such as the stockholders, insurance underwriters, and government regulators, that company directives as well as public policy are being followed. At the same time, these environmental auditors can prove to be very supportive to the line managers. Feedback received from periodic audits will point out areas of strength and weakness and will suggest constructive criticism.

To maintain the independence of the audit department, this group must report directly to senior management through such an office as a vice president for environmental affairs or a vice president for government regulations. The people who staff the audit department should be thoroughly trained in the intricacies of quality assurance, and they should be dedicated to performing their assignments dispassionately. Of necessity they will have to cooperate closely with the operating personnel, but cordial relations should not compromise their objectivity.

Whether audits are announced or impromptu, much valuable insight can be gained. Taking an inventory of labels, laboratory notebooks, manufacturing records, and plant instructions will reveal considerable information about the competence of the personnel. As required, analytical methods may be validated. Plant inspections should help to assess compliance with Good Manufacturing Practice. (The topic of Good Manufacturing Practice is fully discussed in Chapter Five.) Not to be overlooked are reports of customer complaints and their dispositions.[19]

The findings of each audit must be summarized for management and presented with recommendations to the plant. In turn, the plant must respond to the audit report by indicating the corrective actions to be taken. Where differences of opinion exist, these disagreements must be resolved. An effective auditing program thus can be extremely helpful in keeping management informed and simultaneously assisting the plant in doing a better job.

QUALITY CIRCLES

The rationale for quality circles is best explained by the following unpublicized episode. The customer of a supplier of food ingredients reported that a screwdriver had been found in a fifty pound bag of leavening acid. A thorough investigation of the complaint revealed that the foreign object could not have passed through the scalping screen in the filling line and that the only logical way it could have entered the bag before sealing was by someone dropping it inside, either unwittingly or intentionally. No amount of sleuthing, however, could identify the culprit in order to instruct him in better practices or to reprimand him. The chilling conclusion to this investigation was that the entire business was hopelessly at the mercy of a single careless or disgruntled employee.

In this instance management was at a loss where to turn for assistance, but had it organized its operators into quality circles, a built-in device would have been available for recourse. Rather than issuing general warnings and threats of closing the plant, management could have enlisted the goodwill of its work force to find a solution. In a spirit of cooperation, these quality circles could have taken on a grassroots study of the problem. Not unlikely, an imaginative and helpful report would have been forthcoming.

The Structure of Quality Circles

Quality circles are structured as an overlay superimposed on the existing organization. While they can borrow some of the management talent from the present corporate staff, entirely new relationships are created. Figure 1.3 shows in simplified form the key reporting functions. Each circle is comprised of some eight individuals recruited from the rank and file and working in the same area of the operation.[20] The Leader of the circle may be selected from its members or picked from one of the foremen. The Facilitator provides liaison with the rest of the organization, and in addition he is given the assignments of scheduling circle meetings, providing training for members and keeping discussions on course.[21] Overseeing the entire effort is the Steering Committee, with responsibility for formulating objectives and stating policy guidelines.

Once they are presented their charters, circles will define, select, and solve problems encountered in their operations. In tackling these assignments, quality circles are encouraged to use all of the technical, engineering, and managerial skills available. For example, such statistical tools as Pareto charts and scatter diagrams[22] can be used to identify problems. Implementation of findings com-

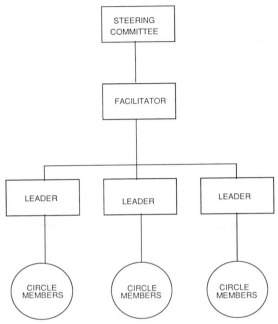

FIGURE 1.3. QUALITY CIRCLES

pletes the process. If participants do not perceive positive results, patience can quickly give way to exasperation. Formal presentations to management help to draw attention to proposed solutions and to get recommendations adopted.

Indoctrination of Members

To succeed, quality circles require a carefully organized program for their introduction. Without the proper indoctrination, workers fail to appreciate the benefits to themselves and therefore are reluctant to participate. One commentator noted that quality circles are not a "Band-Aid for labor problems."[23] Commitment is required from both management and the unions. They must mutually recognize that the "quick-fix" is not the answer, but that measurable gains can only be achieved through a sustained effort.[24]

Many benefits of quality circles have been documented in the literature. Reports of increased productivity, improved efficiency, and better quality have been recounted. First suggested as a management tool in this country, the concept of quality circles made its way to Japan where numerous refinements were introduced. Now this technique is finding new converts in the United States. Prodded not a little by Japanese economic prowess, American business hopes to gain advantage through greater employee participation, enthusiasm, and motivation.[25]

By breaking down the barriers between supervisors and hourly workers, quality circles foster an openness and free exchange of ideas. Instead of foremen talking *to* the operators, management begins to reason *with* its employees. Quality circles promote a feeling that everyone is a part of the same team with a common interest and purpose. Success is sometimes difficult to quantify because results are not always traceable, but experts are unanimous on the necessity of keeping a running tabulation of some meaningful statistics. Not the least significant benefit is the almost certain improvement in worker morale.

CRITICAL PATH METHOD

The Critical Path Method (CPM) and its related management tool, Program Evaluation and Review Technique (PERT) have been used for over twenty-five years in business and government. Management has found CPM and PERT to be powerful aids for planning, scheduling, and controlling the myriad number of jobs which comprise a major undertaking. Initially these techniques found their principal application in the defense industry where projects were large, exceedingly complex and costly, and where performance standards were critical to their success. Later the value of these methods was recognized for a wide variety of assignments including engineering ventures, research and development projects, new product introductions, and military applications, such as missile countdown procedures.

The criteria for selecting CPM (in this treatment no distinction is made between CPM and PERT) are simply that the project must be a one-time event with a definite beginning and end and that it must be made up of discrete activities or jobs. Thus, continuous endeavors or cyclic processes that repeat themselves are not suitable subjects for CPM. By analyzing a complex project in terms of its components, some control can be achieved over its administration. Tradeoffs can be taken between the estimated times for completion and the necessary costs associated with investments, manpower, overtime, and other expenses. The outstanding advantage of CPM is that it eliminates an element of uncertainty, thus helping to avoid unpleasant surprises later on in the form of cost overruns, delays, variances, and outright failure.[26]

Development of a Milk Replacer

To illustrate an application of CPM involving quality assurance, an example is given for the development of a milk replacer. The project is broken down into individual jobs which are arranged in sequential order as listed in Table 1.1. Next to each job is entered its immediate predecessors, i.e., those jobs which must be performed just prior to the one in question, the responsible department for doing the job and the stated time required for its completion. From these data a network diagram is prepared as shown in Figure 1.4. Some experience is needed

TABLE 1.1
JOB LIST FOR THE DEVELOPMENT OF A MILK REPLACER

Job Identification	Job Description	Immediate Predecessor	Department	Time Days
a	Formulation	—	Research	60
b	Process Description	a	Research	10
c	Ingredient Specs	a	Research	20
d	HACCP	b	Manufacturing	20
e	Product Specs	a	Research	15
f	Sampling & Lab Methods	c,d,e	Quality Control	30
g	Kosher Certification	a	Quality Control	5
h	Labeling	g	Labeling	10
i	Packaging	e,h	Packaging	10
j	Ingredient Suppliers	c	Manufacturing	20
k	Suppliers Guaranties	j	Manufacturing	5
l	Product Data Sheets	e	Marketing	20

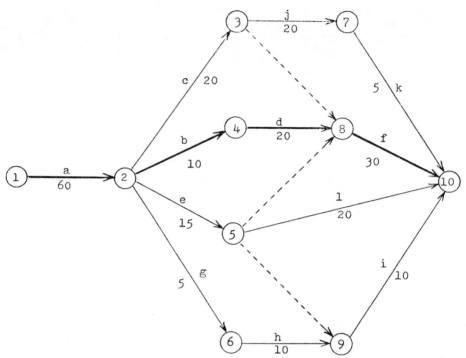

FIGURE 1.4. NETWORK DIAGRAM FOR THE DEVELOPMENT
OF A MILK REPLACER

to draw such a network, but the principles are straightforward. The network arranges the jobs in their proper order and indicates the critical path, or minimum time for overall completion, in bold lines.[27] Software is now available for personal computers that takes much of the drudgery out of using CPM.[28]

To finish the CPM exercise, a calendar can be prepared from the network diagram so that each department knows the timetable to which it must adhere. This schedule, shown in Figure 1.5, indicates the earliest moment when a job can begin and its duration. Only those jobs falling on the critical path, however, must be performed exactly as specified. There is some leeway or slack in the schedules for the other jobs which can be delayed to some extent without jeopardizing the outcome. By preparing and posting a calendar, management puts each member of its organization on notice and renders the departments accountable for their performances.[29]

To date, most applications for CPM have been outside the food industry. There is no reason, however, why wider use cannot be made of this proven method. By expediting new product introductions, where quality assurance is of utmost importance, CPM offers a chance of a quick payoff.

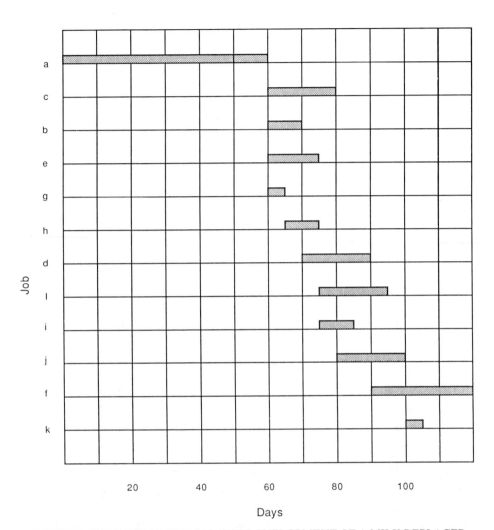

FIGURE 1.5. JOB SCHEDULE FOR THE DEVELOPMENT OF A MILK REPLACER

FOOD AND DRUG ADMINISTRATION

As the lead agency in food regulation, FDA pervades the entire industry. Its operations have a profound effect on all food processors. FDA's success in discharging its duties can be attributed to a strong and dedicated organization, which is structured along functional lines.[30] Where regulatory controversy has arisen, the causes frequently can be traced to a lack of leadership and direction from the executive branch of the federal government.[31]

Early History of FDA

FDA has had a long and tumultuous history. Responsibility for enforcing the first food law was assigned to the Bureau of Chemistry in USDA. Then in 1927 the Food, Drug, and Insecticide Administration, the forerunner of FDA, was established as a separate entity for the first time. The new agency, however, remained under the wing of USDA until 1940. In that year President Franklin D. Roosevelt, exercising his reorganization power, transferred FDA from USDA to the Federal Security Agency. There it remained until 1953 when it was absorbed into a new mega-agency, the Department of Health, Education and Welfare (HEW). Later during the Carter Administration, HEW was stripped of its education role to become the Department of Health and Human Services (HHS).

The wisdom of separating FDA from USDA became apparent during an incident that occurred soon after HEW was organized. FDA had issued guidelines, to go into effect on July 30, 1953, stipulating that wheat, containing rat excreta in excess of one pellet per average pint, be seized as unfit for human consumption. USDA, mindful of the interests of owners possessing large stocks of grain, requested that implementation of the regulation be deferred. A showdown came when Agriculture Secretary Ezra Taft Benson confronted HEW Secretary Hobby in a meeting in full pomp. In spite of the fact that Mrs. Hobby was persuaded to postpone the deadline, FDA's case was forcefully argued. Such an open discussion would not have been likely in the era prior to FDA's independence from USDA.[32]

An Orphan Agency

Although no longer under the dominance of USDA, FDA remains an orphan in search of a permanent home. FDA has been the subject of numerous reorganization plans. One proposal would have split the food and drug functions into two separate agencies, another would have lumped FDA with the Consumer Product Safety Commission, and a third would have tossed the agency back to USDA. The most farsighted proposal was to create a new, independent FDA Commission to be headed by a five member board rather than a single commissioner.[33] This latter idea, however, never received serious consideration in Congress.

An excellent critical analysis of regulatory commissions has been presented by Caspar W. Weinberger, who has held two cabinet posts as Secretary of HEW and Secretary of Defense. The attraction of commissions stems from the basic concept that if a panel of experts is given sufficient independence from the political melee, it will be able to arrive at objective solutions to controversial problems. Beginning with the formation of the Interstate Commerce Commission in the 1880's for the purpose of regulating the sprawling railroads, the number of independent commissions has multiplied so as to constitute a "fourth branch" of the federal government. In Secretary Weinberger's words, Congress originally intended a regulatory commission "to be a sort of deputy of the legislative branch, to deal year-round with day-to-day problems.''[34] The continuing popularity of such commissions, however, testifies to their utility in coping with complex problems in an industrial society.

All of the discussion about reorganizing FDA is not unfounded. Because of their short tenure in office, recent FDA commissioners have failed to provide the continuity of leadership so desperately needed. Furthermore, unlike a board, a single commissioner, acting on his own, is not capable of bringing diverse points of view to bear upon sensitive issues. The problems of managing risk in a free society has been ably discussed by William D. Ruckelshaus, twice appointed administrator of the Environmental Protection Agency. He emphasized the importance of separating the function of risk management from that of risk assessment. When the two assignments become intertwined, the outpouring decisions at best are fuzzy and at worst biased. Those persons responsible for risk management, the policy makers, he said, "need to admit their uncertainties and confront the public with the complex nature of decisions about risk."[35] As future proposals to reform FDA inevitably surface, they should be judged on how well they address the question of risk management.

REFERENCES

1. J. W. Sloat, "Executive's Personal Risk Is High If FDA Cites Unsanitary Storage or Holding," *Food Product Development*, September, 1975, pp. 28-32.
2. Paul M. Hyman, "Criminal Prosecution of Individuals under the Federal Food, Drug and Cosmetic Act," *Food Technology*, May, 1975, pp. 58-62.
3. Mary T. O'Brien and James W. Peters, "Court Ruling Bolsters FDA Prosecution Stance: Top Executives Can Be Held Liable for FD&C Violations," *Food Product Development*, July-August, 1975, p. 20.
4. "Supermarket Head Held Liable for Health Hazard," *The New York Times*, June 10, 1975, p. 22.
5. "FDA Back Change in §701(e) Administrative Procedures," *Food Chemical News*, February 16, 1976, pp. 3-6.
6. Thomas W. Holzinger, "Quality Assurance: National and International," *Journal of Food Protection*, July, 1981, p. 553.
7. Louis J. Bianco, "Guidelines for a Dynamic Quality Control Program in a Changing Market," *Journal of Food Protection*, June, 1977, p. 424.
8. Stephen A. Stockman, "How QC Begat QA at General Foods," *Food Engineering*, June, 1976, p. 55/Int 31.
9. Ben A. Murray, "A Basic Overview of Food Quality Management," *Quality Matters*, November, 1980.
10. C. H. White, "Dean Foods Quality Assurance — Communication and Independence," *Food Product Development*, October, 1978, pp. 40, 41.

11. Morton E. Bader, "Quality Assurance — II, The Quality-Control Laboratory," *Chemical Engineering*, April 7, 1980, p. 89.
12. "Organization Key to Success of Recall Program," *Business Insurance*, February 4, 1974, p. 42.
13. Juliet Bischoff, "Are You Prepared to Manage a Crisis?" *Meat & Poultry*, April, 1987, pp. 22-27.
14. National Canners Association, *Organizing a Product Recall Program*, Bulletin 34-L, revised 1974.
15. Geoffrey Colvin, "Lessons from a Hot Dog Maker's Ordeal," *Fortune*, March 7, 1983, pp. 77, 78, 82.
16. "Humko Gives Quality Control Star Billing," *Chemical Week*, October 7, 1981, p. 44.
17. John A. Byrne, "Are All These Consultants Really Necessary?" *Forbes*, October 10, 1983, p. 144.
18. Wald, Harkrader & Ross, and Resource Planning Corp., *Environmental Audit Handbook*, Executive Enterprises Publications Co., Inc., 1983.
19. The Travelers Insurance Company, Casualty-Property Department, Los Angeles, California, private communication, March 5, 1975.
20. William J. Storck, "Participative Management Brings Employees into Problem Solving," *Chemical and Engineering News*, March 26, 1984, pp. 10-12.
21. "Qualifications for a Facilitator," *IAQC Quality Circle Quarterly*, 3rd Quarter, 1978.
22. Randy England, "Computers for Decision Making," *Chemical Engineering Progress*, September, 1986, pp. 21-23.
23. Perry Pascarella, "Quality Circles, Just Another Management Headache?" *Industry Week*, June 28, 1982, pp. 50-55.
24. "The New Industrial Relations," *Business Week*, May 11, 1981, p. 86.
25. Agis Salpukas, "Quality Circles Aid Productivity," *The New York Times*, May 25, 1981, pp. D1, D3.
26. R. L. Martino, *Finding the Critical Path*, American Management Association, New York, 1964.
27. Jerome D. Wiest and Ferdinand K. Levy, *A Management Guide to PERT/CPM*, 2nd ed., Prentice-Hall, Inc., Englewood Cliffs, New Jersey, 1977.
28. Robert R. Wiggins, "The Best Laid Plans of Mice and Men," *MacUser*, March, 1986, pp. 54-59.
29. Robert W. Miller, *Schedule, Cost and Profit Control with PERT*, McGraw-Hill Book Company, Inc., New York, 1963.
30. "1981 Guide to Government Agencies," *Food Processing*, 1981-82 FP Guide & Directory, pp. 15-19.
31. Rufus E. Miles, Jr., *The Department of Health, Education, and Welfare*, Praeger Publishers, New York, 1974, pp. 240-243.
32. Ibid., pp. 30, 31.

33. "Early Reorganization of FDA Expected under Carter Administration," *Food Chemical News*, November 22, 1976, pp. 65, 66.
34. Caspar W. Weinberger, "Key Features of Government in the U.S.," *Chemical Engineering*, July 17, 1978, pp. 119-122.
35. William D. Ruckelshaus, "Managing Risk in a Free Society," *Princeton Alumni Weekly*, March 7, 1984, pp. 18-22.

CHAPTER TWO
HAZARD ANALYSIS AND CRITICAL CONTROL POINTS

The HACCP acronym is now part of the food technologist's lexicon, but some practitioners still may not understand the origin of the term or fully appreciate its significance. Known to the initiate as Hazard Analysis and Critical Control Points, this topic covers a sophisticated procedure for assessing the dangers inherent in any food process and determining the steps that are necessary to minimize these risks. HACCP is the basis for controlling quality in modern food processes, and as such it forms the underpinning of quality assurance.

BASIC CONCEPTS

For foods to be edible and safe for human consumption, they must conform to approved specifications. It is accepted that these properties are routinely monitored by any food plant, but without supplementary information these tests provide a poor means of controlling an operation. If a product, for example, is found to be out of specification, it is too late for the plant to do anything about its error except to discard, reprocess, or downgrade the material in question for a different end use. Instead of relying on after-the-fact testing of finished product, the plant can avoid this dilemma by monitoring and controlling certain variables within the process.[1] These data will warn the operator of an incipient upset in the process and give him sufficient time to take corrective action before non-conforming product is produced. In a process that is under automatic control, a signal indicating the deviation in the measured variable is fed back to a controller. This feedback which actuates the controller, e.g., a valve, completes the circuit, thus forming a closed loop.

The process variables used to control an operation are identified by a HACCP review.[2] The potential hazards associated with a food process are analyzed, thereby indicating the control points that are critical to a safe operation.[3] Lack of control at any one of these points, e.g., raw material inspection, retorting, or pasteurization, may cause, allow, or contribute to a hazard in the final product.[4] A critical control point has been further defined as a location in the process at which "failure to prevent contamination can be detected by laboratory test with maximum assurance and efficiency.''[5] Because the variables recorded at the critical control points are particularly sensitive to process disturbances, these measurements are extremely useful in anticipating and preventing major problems.[6]

HACCP aids the plant in another way. No processor can afford to sample and test one hundred percent of his finished product. One noted complication is that in some cases accurate data can only be obtained by destructive testing. Even if a processor were inclined to test all of his product, complete protection still would not be assured. Due to unreliable sampling, equipment limitations, human error, or other imperfections, there is always the chance that some inferior product will escape detection. This rule was forcefully illustrated several years ago by a classroom demonstration given in a Short Course sponsored by the Institute of Food Technologists. The students were asked to count the number of times the letter ''a'' appeared in a paragraph set before them. They were given a reasonable but not unlimited amount of time for the exercise. When the answers were collected and reported, the lack of agreement was startling; responses were found to vary from the true value by wide margins.

The food processor can surmount similar shortcomings of finished product testing by establishing in-process controls. Using critical control points he can reduce the frequency of testing and at the same time increase his confidence in the results. No longer does he have to test for everything. He can select those variables which have been found to be particularly significant, and he can cut back the number of analyses to predetermined levels. By eliminating unnecessary sampling, testing, reporting, and record keeping, large savings in manpower and facilities are possible. Improved efficiency can be realized because HACCP permits the plant to allocate its quality control efforts where they will be most productive.

SPACE AGE FEATS

Many of the elements of HACCP have been in use for some time. Only in recent years, though, have the various interrelated parts been meshed together into a unified whole. This development in no small way was an outgrowth of our technological era, and it reflected accomplishments in many disciplines. By tracing the historical events which culminated in the HACCP methodology, a better understanding of its advantages and applications can be gained.

The HACCP concept was born against the background of unimaginable space exploits. In the late nineteen sixties the United States accomplished the visionary goal set before it at the beginning of the decade by President Kennedy. Our nation achieved an unbelievable engineering feat by landing a man on the moon and returning him safely to earth. In the Apollo program no amount of money or effort was spared in an attempt to eliminate all defects. Backup procedures were devised to support those systems most prone to failure. The unprecedented quality assurance program became immortalized by the countdown routine that preceded each launch.

To understand how the National Aeronautics and Space Administration achieved its success, one needs to know something about its approach to quality assurance. NASA has defined quality assurance as:

> A planned and systematic pattern of all actions necessary to provide adequate confidence that the end items will perform satisfactorily in actual operations.[7]

The key word in this definition is "systematic." The interactions of a countless number of components must be understood, and these parts must be put together into a workable entity. Without using a systems approach, no amount of expenditures would suffice to achieve the desired results.

A tragic episode in early 1986 should not detract from the past accomplishments of NASA. When the space shuttle *Challenger* exploded in flight on January 28, killing all seven members of its crew, a general debate ensued over the agency's safety procedures. A Presidential Commission was formed to investigate the cause of the accident and to recommend corrective action. After exhaustive hearings, the commission pinpointed the immediate cause of the catastrophe, namely, malfunctioning O-rings at a joint in the right solid rocket motor. What made the accident all the more regrettable was that, in this particular instance, doubts had been expressed in advance about the integrity of the O-rings.

The commission investigating *Challenger* was also concerned with the broader questions of safety. This concern led one member of the commission, Richard P. Feynman, to theorize about the consequences of ignoring a critical part. Such a part is defined as one which lacks a backup and which, if it fails, will lead to the loss of mission and crew. Professor Feynman said that the launch decision-making was:

> a kind of Russian roulette. . . . [The shuttle] flies [with O-ring erosion] and nothing happens. Then it is suggested, therefore, that the risk is no longer so high for the next flights. We can lower our standards a little bit because we got away with it last time. . . . You got away with it, but it shouldn't be done over and over again like that.[8]

The theme expressed by Professor Feynman is that, contrary to conventional wisdom, with each completed mission by the shuttle, its future chances of success *decreased* rather than improved. The disregard for the reliability of the O-ring critical part inevitably led to the shuttle's doom. In this respect, a critical part is to a space vehicle what a critical control point is to a food process.

Natick Laboratories' Program

While NASA was setting the stage for its space feats, a parallel effort was underway at the U.S. Army Natick Laboratories in Massachusetts to perfect a zero-defects program for the food industry. This project had merit equal to our space endeavors since where the public health is involved, no compromise with perfection is tolerable. The Natick Laboratories succeeded in developing a system, known as Modes of Failure, which examined a given food product, each of its ingredients, as well as the process conditions necessary to prepare it. At each step the question was asked, what can possibly go wrong with the total system? The work performed at Natick was an important forerunner to HACCP.[9]

Some people might wonder about any analogy drawn between the space program and the food industry. Indeed, until recent times they had little in common, but because of the transformations that have taken place in the food industry the two fields are now similar in their complexity and their dependence on technology. The changes have been prompted by consumers who not only expect wholesome and nutritious food at reasonable prices but demand, as never before, infinite variety, convenience and palatability. To keep abreast of the challenges, food processors have rapidly assimilated new advances in science, including the concepts spun off by the space program. Thus, today the food industry is virtually on the leading edge of technology. Completing the cycle, the food industry is proud of the role that it has played in creating completely new and unique products to sustain the astronauts on their journeys.

DENVER CONFERENCE

Progress in the food industry has not always been steady in spite of its recent commendable record. For many years producers resisted outside efforts to reform their businesses, preferring instead to stick with methods that were familiar to them. This state of affairs was documented in a report issued to Congress on April 18, 1972, by the Comptroller General of the United States. It was titled "Dimensions of Insanitary Conditions in the Food Manufacturing Industry," and it covered a survey of 97 plants. The findings indicated a deplorable lack of standards: by General Accounting Office criteria, 40 percent of the plants inspected were considered to be insanitary, or by the Food and Drug Administration's guidelines, 23 percent were placed in this category. The impressions of the report were summarized as follows:

> During the past three years, FDA inspections have indicated that sanitary conditions in the food industry in the United States are deteriorating. FDA did not know how extensive these insanitary conditions were and therefore could not provide the assurance of consumer protection required by the law. A serious problem of insanitary conditions exists in the food manufacturing industry. Several actions must be taken by FDA to alleviate these conditions.[10]

1971 National Conference on Food Protection

Responding to public pressure for improved conditions in the food industry, the American Public Health Association convened a conference in Denver on April

4, 1971. Billed as the 1971 National Conference on Food Protection, this meeting addressed many topics of food safety. One of the workshops was assigned the subject, Prevention of Contamination of Commercially Processed Foods. After a thorough review of the problem this panel put forth the idea of focusing attention on the critical control points in a food process. It concluded that the more effective and efficient approach to prevent contamination was to concentrate attention on the most vulnerable or critical steps of a process rather than to spread limited resources across the board. In his closing remarks to the conference, Howard Bauman, Vice President of The Pillsbury Company, noted with insight:

> The concept of hazard analysis of each food and food system and the establishment of critical control points to ensure quality was presented and recommended for widespread use.[11]

HACCP Procedure

The core idea of Hazard Analysis and Critical Control Points, introduced at the Denver Conference, quickly took root in the food industry.[12] When Congress considered new food legislation in 1975, a key provision of Senate bill, S. 641, provided that:

> . . . a food processor shall —
> A. Identify those control points in the food processing operation carried out by any establishment owned or operated by him which are important in the prevention of adulteration, within the meaning of Section 402(A) of this chapter;
> B. Identify the hazards associated with each such point;
> C. Establish adequate controls at each such point; and
> D. Establish adequate monitoring of the controls at each such point.

Although no action has been taken to date by Congress on these proposals, this inertia in no way reflects on the validity of HACCP. In the meantime FDA has instituted routine HACCP inspections, which it considers to be far more productive notwithstanding the considerable extra time and training needed by inspectors.[13] Industry has also picked up the theme. In 1975 the Association for Dressing and Sauces developed model guidelines for its members to assist them in "developing their own critical control point program, to identify hazards at these points, establish adequate control and monitor such controls."[14]

MICROBIOLOGICAL HAZARDS

The hazards which plague food processors are legion. For convenience they are grouped into categories as shown in Table 2.1. A further advantage of listing these hazards by type is that this grouping helps in their identification and suggests means for their control. No compilation of this kind will be exhaustive; it is practically impossible not to overlook some mentioned dangers, and new

TABLE 2.1
FOOD HAZARDS

A. BACTERIA	H. RADIOACTIVE ISOTOPES
Clostridium botulinum	Cesium 137
Clostridium perfringens	Iodine 131
Salmonella	Potassium 40
Staphylococcus aureus	Strontium 90
B. MOLDS	I. EXTRANEOUS MATTER
Aspergillus flavus	Filth
Penicillium cyclopium	Glass Splinters
C. PARASITES	Peeling paint
Tape worms	Tramp metal
Trichinellae	J. NATURALLY OCCURRING TOXINS
D. PESTS	Ciguatera poisoning
Birds	Oilseed toxins
Insects	K. NUTRITIONAL DEFICIENCIES
Rodents	Infant formula exceptions
E. RESIDUES	Processed foods
Antibiotics	L. REGULATORY HAZARDS
Chlorinated insecticides	Label errors
Organophosphate insecticides	Short weights
F. INDUSTRIAL CHEMICALS	M. FUNCTIONAL HAZARDS
Hexabromobiphenyl	Packaging defects
Polychlorinated biphenyls	Particle size deviations
Vinyl chloride	
G. HEAVY METALS	
Arsenic	
Cadmium	
Lead	
Mercury	
Selenium	

risks are being revealed daily. Not all reported hazards constitute a real threat. There are many false, misleading, and emotional reports. One area where feelings are running high is the subject of food additives, several of which are said to be carcinogenic or otherwise deleterious. In Chapter Three, a discussion of these substances attempts to put this topic in perspective. The conscientious food technologist will remain attentive to each new disclosure, reserving judgement until substantial evidence is available. Various sources of information are available to him, such as the *Morbidity and Mortality Weekly Report* issued by the Centers for Disease Control of the Public Health Service, U.S. Department of Health and Human Services. This periodical plus the notices of recalls posted by the Food and Drug Administration provide timely warnings.

Bacterial Food Poisoning

Bacteria lead the list of hazards in terms of their widespread occurrence and the dangers which they pose. Microbes are ubiquitous in nature, and given the right conditions for growth, they will multiply exponentially. For example, in a favorable environment *Escherichia coli* will approximately double its population every twenty minutes.[15] Many of our more nutritious foods such as milk, eggs, meat, fish, and soup are ideal media for bacterial growth. When held at warm temperatures, in the range of 45 °F (7.2 °C) to 140 °F (60.0 °C), these foods meet the environmental needs for the rapid reproduction of many microorganisms.

Food poisoning is commonly of two types, food intoxication where a bacteria produced toxin induces an illness, and food-borne infection which results from exposure to live pathogens. *Clostridium botulinum* is responsible for the most feared form of food poisoning. This microbe under low-acid, anaerobic conditions will produce a toxin that is one of the most poisonous substances known. Even in minute quantities it will cause sudden death. Control of *C. botulinum* is made difficult because it forms spores that are heat resistant. Any viable spores remaining in improperly canned foods will germinate and multiply, producing their deadly toxin. Although the toxin is heat-labile and can be made innocuous by boiling, in all instances where botulism is suspected, the food should be discarded.

Another food intoxication is caused by *Staphylococcus aureus*. People are carriers of this pathogen which occurs in boils, skin lesions and nasal passages. Under insanitary conditions food can be contaminated with these organisms which will then multiply in the host environment under favorable temperatures. Typical foods which can be adulterated are pastries as well as hams and other cured meats since the organism is tolerant to salt and nitrates/nitrites. *S. aureus* produces a powerful exotoxin that causes vomiting, cramps, and diarrhea. Because the toxin is heat stable and may be present after the organism is destroyed, diagnosis of the illness can be complicated. In the past, numerous suspected cases of intoxication probably have been attributed to unknown causes.

Food-borne infections are frequently caused by *Clostridium perfringens* and by different serotypes of *Salmonella*. The former organism is responsible for many outbreaks of food poisoning in the food service industry, particularly at catered events. It is most common when food is prepared in advance and kept warm for hours before serving. *C. perfringens* is an anaerobic, spore-forming bacteria that requires certain substrates such as meat, poultry, or dressings and gravies made from these foods. Contamination may be caused by poor personal hygiene of the food handlers. With symptoms of nausea, intestinal gas, and diarrhea, the sickness is discomforting but not especially serious. The other infection, *Salmonella*, is transmitted primarily by farm animals which pass the organisms on to such foods as eggs and meat products. People infected by this disease experience a range of symptoms from diarrhea, nausea and abdominal pain to chills, fever, and vomiting. In spite of concerted efforts by industry and government agencies to stamp out *Salmonella*, this pathogen remains prevalent in our food supply.

Unexpectedly, a relatively new bacterial infection has made the news. *Listeria monocytogenes* was the cause of 39 deaths in California as the result of the contamination of Mexican-style soft cheese during June, 1985. Produced by Jalisco Mexican Products Inc., now out of business, this cheese was mishandled under insanitary conditions at the company's Artesia plant.[16,17] Concerted efforts to categorize this pathogen reveal that it has an uncanny ability to survive under adverse conditions and that it is able to grow at refrigeration temperatures. For these reasons, such soft cheeses as Brie and Liederkranz are particularly susceptible.[18]

Indicator Organisms

Non-pathogens are a concern in food sanitation even though many benign organisms like lactobacilli and yeast are purposefully employed to make such products as yogurt, sauerkraut, wine, and bread. Food spoilage in large measure is due to the uncontrolled contamination by microorganisms. As food becomes putrid, ptomaine poisons may be formed. For example, when scombroid fish decomposes, free histidine is decarboxylated by bacteria to histamine. Bacteria levels in food are of interest for yet another reason. Their presence in excessive numbers indicates a lack of cleanliness and therefore possible contamination by pathogens. For this reason "indicator" organisms are measured by such tests as Aerobic Plate Count, Indicator Count and the *E. coli* count. High counts in the latter two tests implicate human fecal pollution. Specifications for these and other indicator organisms are routinely set by suppliers of perishable or sensitive food ingredients.

Molds

Molds are receiving increased attention as more data are being accumulated about them. These microorganisms not only account for large food losses but frequently produce mycotoxins. *Aspergillus flavus*, a mold found on cereal grains, has proved to be especially troublesome for peanut growers. This mold produces a potent mycotoxin called aflatoxin that has been found to be carcinogenic. Many molds commonly occur in the home kitchen where they are known by their appearance, if not their taxonomy, to housewives. An example of one of these molds is *Penicillium cyclopium* which often is found on stored foods and produces a mycotoxin. Care should be taken to avoid eating molds except such intentional ones as veined in Roquefort and other blue cheeses.[19]

Parasites and Pests

Parasites and pests are two more classes of food hazards. Parasites are represented by a variety of tape worms as well as by trichinellae roundworms which may be obtained by eating undercooked pork. Except possibly for trichinosis, parasites no longer are a major problem in the United States, but when they do occur they are debilitating and difficult to treat.[20] Pests including birds, rodents and insects, are the most obvious signs of insanitary conditions in a food plant, warehouse, or retail establishment, and for this reason they receive a disproportionate amount of attention. Nevertheless vermin are a menace, primarily because they are vectors of disease, and care must be taken to exclude them from these premises. Insect fragments and rodent hairs found in raw foodstuffs indicate possible adulteration. Quality control personnel have found ultraviolet radiation, known as "black light," to be effective in detecting the presence of rodents and mold spoilage. Body fluids from rodents and many molds are distinctly fluorescent under uv light.

CHEMICAL CONTAMINATION

In our modern industrial society chemical contamination is a growing concern to the safety of our food supply. Two noted sources of these contaminants are the residues of pesticides used on crops and the traces of antibiotics which have been added to animal feeds. In addition, there has been an increasing background of synthetic chemicals that have pervaded our environment. The problem becomes acute when, on occasion, a serious chemical spill takes place. The regulatory authorities have been handicapped by the lack of personnel to supervise this activity. Therefore priorities for the surveillance of chemicals have been set by such considerations as the following:

1. Volume of production. The greater the production, the more likely the contamination.
2. Toxicity of the chemical.
3. Solubility behavior. A fat soluble material is more likely to bioaccumulate in food animals.
4. Environmental stability.
5. End uses. Products used in "open" applications are more apt to contaminate foods than those used in closed systems.
6. Means of disposal. Products disposed of properly are less apt to contaminate than those just "dumped."[21]

Two incidents, which grabbed front page headlines in the 1970's, illustrated the need for closer surveillance of chemical substances. In late 1970 polybrominated biphenyls (PBB), a fire retardant manufactured by Michigan Chemical Co., inadvertently was blended into animal feed rations and consumed by thousands of cattle, hogs, sheep and chickens throughout the State of Michigan. Before this disaster, known as "Cattlegate" in the press, was over, a large segment of the state's population had been exposed to PBB and more than thirty thousand cattle and millions of eggs had to be destroyed. Many complaints of illness were linked to this contamination. An investigation later revealed the likely cause for the error: poorly labeled containers of PBB were mistakenly used in place of the intended feed supplement, magnesium oxide.[22]

In another major blunder that happened in June, 1979, a polychlorinated biphenyl (PCB) spill spread its effects over a nineteen state region, as well as Japan and Canada. The accident began when a forklift operator at the Pierce Packing Co. plant in Billings, Montana, ran his truck into a stored electrical transformer which contained PCB. The fluid leaked into a drainage system that led to the rendering operation. As a consequence, close to two million pounds of feed were adulterated. Much of this product was consumed before being detected, and therefore millions of eggs plus untold quantities of pork had to be destroyed. This catastrophe taxed the resources of state and federal agencies, and it resulted in losses that exceeded $10 million.[23]

Heavy Metals

Heavy metals are considered in a separate hazard category from other chemicals because of their pernicious nature. Arsenic, lead, mercury, and selenium attack the central nervous system, and in larger doses they cause fatality. The body's defenses against heavy metals are limited; lead, for example, is a cumulative poison that builds up in the body until symptoms and disability occur.[24] Food Chemicals Codex, (this is a standard reference which is discussed in Chapter Three) recognizing that many minerals and natural products are possible sources of heavy metals, sets limits for these impurities, notably arsenic and lead. Another source of lead is the solder used to seal tin cans. Measures have been taken in recent years to reduce the lead contamination of baby formulas, particularly condensed milk, as well as juices and other adult foods preserved in cans. Fish and seafood taken from polluted waters are major contributors of mercury poisoning. Under the auspices of the United Nations Environment Program, plans were announced in 1978 to clean up the Mediterranean Sea in order to reduce the levels of mercury and other heavy metals.[25] Cadmium finds many industrial uses, one of which is to plate steel products in order to provide better rust prevention than afforded by galvanization. Nevertheless Sweden's Product Control Board in 1979 recommended a partial ban on the use of cadmium with the aim of reducing the amount of this metal found in farm produce.[26] Other steps to minimize cadmium pollution were proposed in 1979 by the Commission of the European Communities. This organization advocated reductions in the cadmium content permitted in sewer sludges that are spread on crop lands.[27]

Radioactive Isotopes

Radioactivity is a phenomenon of the atomic age. Although there has always been a background of natural radioactivity, this level is insignificant compared with recent contributions made by man. Until the superpowers negotiated a ban on the atmospheric testing of nuclear weapons, significant quantities of cesium

137 and strontium 90 found their way into the milk supply. Now the chief concern is such accidents as the nuclear reactor malfunction at Three Mile Island, Pennsylvania, in 1979[28] and the 1986 nuclear disaster at Chernobyl in the Ukraine, Russia.[29]

Extraneous Matter

The danger of introducing extraneous matter into food seems too obvious to belabor, and yet this hazard leads to very substantial losses. Scheduled maintenance notwithstanding, food plants seem to be constantly falling apart. Loose bolts, broken sieves, flaking rust, peeling paint, glass fragments, and wooden splinters are more common than most operators would care to admit. Other foreign matter like stones and bones may be introduced with the raw ingredients. Filth is a catchall class that includes whatever comes along.

Naturally Occurring Toxins

Naturally occurring toxins abound. Everyone has heard of poisonous mushrooms, but few are knowledgeable enough to distinguish the good from the bad varieties. With the popularity of fish and chips, one may wonder about the edibility of all the fish caught by oceangoing trawlers. Numerous reports of ciguatera poisoning are attributed to fish which have been feeding on natural toxins.[30,31] A final example of natural toxins involves oilseeds which for many years have been fed to livestock after the oil has been expressed. Because of the increased interest in upgrading the nutritious protein values contained in these seeds, attention has been focused on ways to condition such oilseeds for human consumption. Before cottonseed is suitable to eat, a natural toxin, gossypol, must be extracted or destroyed to meet safe levels.

Nutritional Deficiencies

Nutritional deficiencies of processed foods are of great concern because of the accelerated introductions of these products. These so-called engineered foods lack many of the vitamins, minerals, and other nutrients needed for a balanced diet. Unless these deficiencies are restored, people risk the danger of malnutrition. The substitution of margarine for butter in many diets is a case in point. Because vegetable oils, unlike butter fat, do not contain vitamins A and D, fortification of margarine is necessary to replace the lost nutrients. There is no greater concern in nutrition than the assurance of safe baby foods. In response to this priority Congress passed special legislation for infant formulas, a subject which is reviewed more fully in Chapter 5.

Regulatory and Functional Hazards

Regulatory and functional hazards cover a range of problems that beset the food processor. Labeling errors can lead directly to the misuse of a product. Take, for example, a vitamin blend that is designed for flour enrichment. If the wrong label is affixed to the container of the blend, the resulting flour will contain incorrect amounts of the added nutrients. Short packaged weights not only will cheat a customer of product, but in cases where ingredients are consumed on the basis of so many containers of each raw material, incorrect formulations may result. Packaging defects present all kinds of potential difficulties ranging from product contamination to poor sanitation. An entire chapter is devoted to this subject later in the book. Finally, lumps and caking can render a product unusable. The sieve analysis of an ingredient can be vital to its proper application.

FOOD PRESERVATION

Mankind has learned to cope remarkably well with the many hazards as described in the previous sections. If he had not succeeded in these endeavors, civilization would not have evolved as we know it. Man's most insidious enemies in his struggle for survival are microorganisms, which in one unguarded moment can ravage a food supply. Through resourcefulness and perseverance, man early on found out how to prevent food from spoiling by means of fermentation, pickling, brandying, salting, preserving, drying, and smoking. More recent methods include canning, pasteurization, and freezing while irradiation, which is just beginning to be sanctioned on a selective basis, offers considerable hope for the advancement of food preservation.

Canning

The history of canning goes back to the year 1795 when in response to military needs the French government offered a 12,000 fr. prize for the discovery of a practical food preservation method. Motivated by this offer, Nicolas Appert in 1809 succeeded in demonstrating the canning of foods by sealing them in glass containers and heating or retorting the filled jars for a given period of time. Without any knowledge of the theory underlying food preservation, progress in canning depended on trial and error to determine the necessary processing conditions. Unfortunately the results were not always reliable. Not until around 1860 when Louis Pasteur's epic discoveries laid the foundation of microbiology did the principles of canning become understood. Progress was furthered in the United States by the brilliant team of Samuel C. Prescott, a professor at Massachusetts Institute of Technology, and William L. Underwood, a food industry leader. Working together in 1895 they tackled and solved the vexing problem of canning corn. Now safety of all low-acid canned foods is assured by strict government regulations which are based on the pioneering efforts of these and other workers.[32]

Pasteurization

Pasteurization is a well established practice that was developed by Louis Pasteur for preventing spoilage in beer and wine. The process involves a mild heat treatment for a specified length of time; the higher the temperature applied, the shorter the time required. In the conventional process for pasteurizing milk, every particle of the milk must be heated to at least 145° F (62.8°C) and held at this temperature for a minimum of 30 minutes. Alternatively the Pasteurized Milk Ordinance (see Chapter 5) permits the use of High-Temperature, Short-Time (HTST) methods[33], which have been shown to be equally effective in destroying pathogens and at the same time advantageous in developing less off-flavor. The lower limits of time and temperature are based on the need to destroy the disease-causing *Mycobacterium tuberculosis*. These conditions will also control *S. aureus*. Ultra High Temperature (UHT) pasteurization is just being introduced into the United States after having demonstrated its value in European countries. This process, which sterilizes the product, is euphemistically described by the statement, "aseptically processed and packaged." Milk products so treated have extended keeping properties at room temperature and therefore do not require refrigeration.[34]

Freezing

The commercial development of freezing dates back to 1842 when an English patent was granted for freezing such foods as meat and fish in brine. Not until the advent of mechanical refrigeration, however, did freezing assume significant proportions. The first shipment of frozen beef from Australia to Great Britain was made around 1880. In spite of early progress made with freezing, this process was not adaptable to fruits and vegetables. The cell structures of these foods are broken down by conventional freezing methods so that on thawing, the firmness and crispness of the fresh produce is lost. The problem was overcome by the outstanding research of Clarence Birdseye working during the 1920's with General Foods Corporation. His efforts resulted in the introduction of small containers of frozen food for the retail trade. By subjecting vegetables to a blanching step to inactivate the enzymes present and then flash freezing the product, much of the original flavor, appearance, texture, and nutritive value is retained. Today the consumer enjoys a wide variety of frozen foods, from orange juice concentrate to TV dinners.

Irradiation

Food irradiation deserves special mention because of its promise for the future. Exposing food to x rays or to gamma rays from cesium 137 or cobalt 60 has been found to be effective in pasteurization, sterilization, mold inhibition, and the prevention of sprouting.[35] Anything suggestive of radiation has engendered emotional reactions from consumers. So far these fears have proven to be groundless as extensive toxicological studies have failed to uncover the slightest peril. In 1953 the U.S. Army Natick Research and Development Command began the investigation of irradiation, and it pursued this activity vigorously until the program was transferred to the U.S. Department of Agriculture in 1980. International studies have been coordinated by the Food and Agriculture Organization/International Atomic Energy Agency/World Health Organization (FAO/IAEA/ WHO) Joint Expert Committee on Food Irradiation (JECFI) which first began meeting in 1964.

Although irradiation is not recommended for dairy products because of organoleptic problems, many other applications have been tried with good results. The irradiation of strawberries has been shown to prevent mold growth which destroys an estimated 25 percent of the post harvest crop.[36] Low level irradiation of white potatoes will inhibit sprouting during storage. Low doses will also effectively disinfest grains and spices and thus can replace the controversial fumigants, ethylene dibromide and ethylene oxide, which are reported carcinogens. There have been several independent studies of the use of irradiation of fish and seafood to reduce bacteria counts and thus improve their quality.

The shelf life of ground beef can be significantly lengthened by using irradiation to destroy much of the microflora, particularly the pseudomonas which are slime-forming, putrefying bacteria. Without benefit of these facts but in response to a public outcry for better meat, Oregon, in a major departure, established microbial standards for these products in 1971. These specifications were tightened in October, 1973, as follows:

1. Meat food products, whether fresh, frozen, prepared, or otherwise manufactured, shall be deemed to be adulterated . . . if:
   (a) The microbiological level exceeds 5 million organisms per gram in fresh or frozen meat food products (including ground, chopped, fabricated, and whole cuts thereof), or exceeds 1 million organisms per gram in meat food products which have been cooked, smoked or otherwise heat treated; or
   (b) The most probable number (MPN) of E. coli organisms exceeds 50 per gram in fresh or frozen meat products (including ground, chopped, fabricated, and whole cuts thereof), or exceeds 10 per gram in meat food products which have been cooked, smoked, or otherwise heat-treated.
2. The provisions . . . shall not apply to meat food products which have been fermented or inoculated as a procedure of preparation or manufacture.[37]

The state subsequently found that with the practices that then existed in the meat industry, many suppliers were not able to comply with this regulation.[38] In all probability, new technology encompassing irradiation could meet these standards with considerable advantage to consumers.

Irradiation is generally classified as low dose when applied below 1 kiloGray (kGy), medium dose between 1 kGy and 10 kGy, and high dose over 10 kGy. A number of parameters are critical in successfully applying irradiation: the dose, dose rate, temperature, water concentration, and oxygen level. Experimental data have shown that the kill fraction is dependent on the dose, whereas secondary effects, such as the formation of free radicals, are influenced by the dose rate.[39] Thus, by increasing the dose rate and holding the dose constant, much the same results can be obtained as with HTST pasteurization: effective biological inactivation is achieved with the development of less off-flavor. At high doses, sterilization can be effected by irradiation. Unlike retorting where possible "cold spots" present the danger of underprocessing, irradiation energy is distributed quite uniformly.[40] Complete protection can therefore be assured. Irradiation has another advantage because it is suited to the preservation of solid or chunky foods, whereas canning requires the food to be packed in a liquid such as juice, gravy, or a sauce in order to achieve adequate heat transfer during retorting. Proof of the effectiveness of irradiation in food preservation was demonstrated by serving meats so processed on the historic flights of Apollo 17, Apollo-Soyuz, and the Space Shuttle.

Convinced about the safety of irradiation, FDA in April, 1986, approved the application of low doses to inhibit the growth and maturation of fresh fruits and vegetables and to control insects. At the same time, the agency permitted increased doses, up to 30 kGy, to disinfest spices and herbs of microorganisms. In addition, low level irradiation, 0.3 kGy to 1 kGy, has been permitted by USDA for the treatment of fresh pork to reduce the risk of trichinosis. Prior to these clearances, FDA had sanctioned irradiation to control insects in wheat and flour as well as to prevent sprouting of potatoes. Foods treated with radiation must display the accepted international logo plus the statement "treated with radiation" or "treated by irradiation." After two years experience with commercially treated food products, the radiation statement may be dropped at the discretion of FDA.[41,42]

CRITICAL CONTROL POINTS

When everything is critical, nothing is critical. If all of the control points in a process are assumed to be critical, they, de facto, lose their cardinal importance. Therefore the selection of more than one or at most two critical control points for each hazard is self-defeating. In agreement with the definition for a critical control point, the manufacturer should single out for special attention that location in a process which offers the best opportunity to detect imminent trouble.

The identification of the critical control points requires a thorough familiarity with the food process, including each step in the operation from the receiving of raw materials to the shipping of finished product. A flow diagram showing all of the processing streams is indispensable for understanding the key elements of the process. With this knowledge a systematic search can then be made to identify the potential entry points of each hazard.[43] Many of these entry locations are obvious, e.g., raw materials, but others are less apparent, such as open vats, exposed conveyor lines, and even the equipment itself. The latter may be a source of hazards because of broken parts or unclean surfaces. It is axiomatic that all food processes are open systems which are vulnerable to external effects that one way or another must be controlled.

The following observation of a food microbiologist testifies to the need for total control over all potential entry points in a food process.

> Much has been said about the "ubiquity" of the Salmonella bug. I do not prescribe to this theory. If the Salmonella bug is *there*, it must come from *somewhere*. The answer is a total program of continuous regular sampling of ingredients and product, and consideration of environment. The environment includes anything that might eventually end up in product, either through direct or indirect contact. For example, in chocolate production, scrap chocolate is a good indicator; in dry milk, sifter tailings, residues from vacuum cleaners, and shoe scrapings; in dehydrated foods, air filters and air.[44]

Blended Products

Many food products, such as cake mixes, are a composite reflection of their raw materials. Such properties as nutrient levels, microbe contamination, and percentages of impurities are additive and depend solely on the quality of the ingredients. Under such circumstances, the critical control points for these attributes are located at the raw material stage. An example of such a process is shown in Figure 2.1. It shows a schematic diagram for the preparation of a milk replacer. The critical control points are indicated by numbered diamonds and the other control points by numbered circles. To illustrate the difference between them, control point No. 2 might test for the protein content of the soy flour while critical control point No. 2 would test for *Salmonella*. Critical control points 4, 6, and 7 are established to check for tramp metal and other foreign objects. Critical control point 5 monitors product uniformity while No. 8 is for final product testing. Finally, critical control point 9 records the filled weights of the containers.

From the above example, one will observe that not every raw material presents the same degree of risk. While microbiological tests were specified for the whey and soy flour, no such controls were needed for the calcium carbonate. Reflecting such differences, food ingredients can be arranged in different categories on the basis of their sensitivity to microbial contamination. Table 2.2 shows but a few of such ingredients. Category I is reserved for those ingredients requiring extra care because of their special applications, namely infant formulas and diets for the infirm. Ingredients which are easily susceptible to contamination, such as eggs and milk, are grouped into Category II. Less sensitive products, including

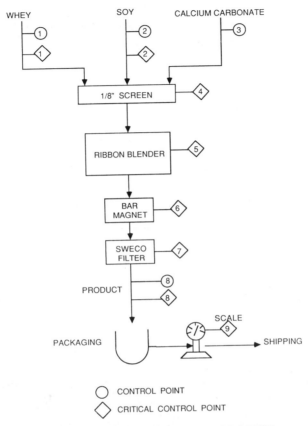

FIGURE 2.1. MILK REPLACER PROCESS

flour, soy, starch, and shortening, are put in Category III. Lastly, Category IV is designated for all ingredients that historically have been found to be free of pathogens. Table 2.2 can be expanded to provide a ready reference to those concerned with the identification of hazards.

Susceptible Ingredients

In contrast to the blending of such products as milk replacers, many food processes must take for granted the microbial contamination of its raw materials. In these situations the processor has to depend on a "kill step" to eliminate or control potentially harmful organisms.[45] The definition of a kill step is a process adjustment which will achieve the reproductive inactivation of a given microorganism. This critical control point may consist of a heat treatment, an acidification step, an adjustment in water activity, or the addition of an antimicrobial agent. In the processes just described in the preceding section, heat treatments are

TABLE 2.2
SENSITIVITY CATEGORIES FOR FOOD INGREDIENTS

I. INGREDIENTS FOR SPECIAL GROUPS
   INFANTS
   INFIRM
   GERIATRIC

II. SUSCEPTIBLE INGREDIENTS
   DAIRY PRODUCTS
   EGGS
   FISH
   MEAT
   SOUP

III. INSENSITIVE INGREDIENTS
   FLOUR
   SHORTENING
   SOY
   STARCH

IV. INGREDIENTS FREE OF PATHOGENS
   BAKING SODA
   CITRIC ACID
   FOOD COLORS
   MONOSODIUM GLUTAMATE
   SALT
   SUGAR

employed in canning and pasteurization. The reverse, namely cooling, is used as a control in freezing. In food irradiation processes, the kill step is the radiation itself. Pickling depends on acidification to stop bacterial growth while many fermentation processes rely on lactic acid forming bacteria to lower the pH. Other fermentation processes use yeast to produce ethyl alcohol, which acts as a bacteriostat in brandied foods. Water activity is adjusted by such methods as drying, salting, and preserving with sugar. Microbicides are released in smoking, or they may be supplied by such additives as sulfites, benzoates, and propionates.

Microbial Safe Ingredients

To complete the discussion of critical control points, some comments should be made about microbial safe ingredients such as citric acid, baking soda, monosodium glutamate, and certified food colors. Because these products are essentially free of microorganisms, tests for these cells can be ignored. On the other hand tight specifications are established for the purity of these compounds as a means of limiting undesirable or harmful constituents. An example of such an impurity is the oxalic acid produced as a by-product in the fermentation process for citric acid.

Purification steps are invariably employed in processes for the manufacture of food ingredients. Phosphoric acid is treated with a sulfide to remove heavy metals. Frequently a crystallization operation is used in the final step in the production of a salt. In such a case the purity of the solution fed to the crystallization step becomes a critical control point. If the assay of the final product must be equal to 99 percent, then the feed liquor to the crystallizer might require a purity, for instance, between 80 and 85 percent. The higher the purity within this range, the larger the crop of crystals that can be obtained and still meet the product specification. By making such adjustments, very close control over the process can be realized. Figure 2.2 indicates the location of the critical control points for controlling purity in the manufacture of monosodium glutamate via fermentation.

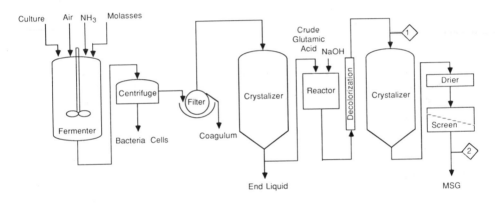

FIGURE 2.2. MONOSODIUM GLUTAMATE (MSG) PROCESS

Bon Vivant Scandal

What happens when control is lost over a critical control point? To everyone's dismay and horror the consequences of such an irregularity were vividly narrated in a 1971 front page story. In the heat of summer a New York commuter and his wife opened a can of Bon Vivant vichyssoise for supper. By the next day he was dead from botulism, and his wife was suffering from symptoms of paralysis. An investigation of Bon Vivant's Newark, New Jersey, soup plant by the Food and Drug Administration revealed "totally inadequate and incorrect manufacturing and record-keeping procedures." Failure to insure adequate thermal processing at the critical retort step exposed the operation to the most serious hazard known in the food industry. Even though a general recall of all suspected product was undertaken and no other accounts of illness were reported, the soup company, a family business going back four generations, never recovered from this tragedy. Bon Vivant, a name which had been associated with premium soups, became overnight a byword for food processing malpractice.[46]

REFERENCES

 1. Riemann and Bryan, *Food-Borne Infections and Intoxications*, 2nd ed., Academic Press, 1979, p. 697.
 2. Virgil O. Wodicka, "The Food Regulatory Agencies and Industrial Quality Control," *Food Technology*, October, 1973, p. 58.
 3. R. Angelotti, "FDA's Plan for Quality Assurance in the Food Industry," *Food Product Development*, December, 1975, p. 14.
 4. F. Leo Kauffman, "Interpretation and Application of FDA Laws and Regulations," a paper presented at a joint meeting of the Food Processors Institute and the Institute of Food Technologists, Chicago, Ill., October 13, 1975.
 5. American Public Health Association, *Proceedings of the 1971 National Conference on Food Protection*, U.S. Department of Health, Education, and Welfare, 1971.
 6. Joseph P. Hile, "HACCP — A New Approach to FDA Inspections," a paper presented to the Food & Drug Law Institute, Washington, D.C., December 11, 1973.
 7. R. G. Landis, "What Is Industry's Approach to Quality Assurance?" *Food Technology*, October, 1970, pp. 48-50.
 8. *Report of the Presidential Commission on the Space Shuttle Challenger Accident*, No. 040-000-00496-3, Government Printing Office, Washington, D.C. 20401, June 6, 1986, p. 148.
 9. Howard E. Bauman, "The HACCP Concept and Microbiological Hazard Categories," *Food Technology*, September, 1974, p. 30.
10. Robert W. Harkins, "Striking a Balance — Industry Responsiveness and Regulatory Control," *Food Product Development*, February, 1974, p. 42.
11. American Public Health Association, *Proceedings of the 1971 . . .* , op. cit.
12. Mary T. O'Brien, "Self-Generated Safety Assurance Dictated by S. 641," *Food Product Development*, December, 1975, p. 68.
13. Virgil O. Wodicka, "The Food Regulatory Agencies and Industrial Quality Control," *Food Technology*, October, 1973, p. 58.
14. *Food Chemical News*, November 24, 1975, p. 13.
15. Roger Y. Stanier, Michael Doudoroff, Edward A. Adelberg, *The Microbial World*, Prentice-Hall, Inc., Englewood Cliffs, N. J., 1957.
16. "Cheese Is Linked to 44th Death," *The New York Times*, June 25, 1985, p. A20.
17. "2 Officials Accused over Tainted Cheese," *The New York Times*, March 28, 1986, p. A16.

18. "New Bacteria in the News," *Food Technology*, August, 1986, pp. 16-26.
19. Harold Hopkins, "Danger Lurks Among the Molds," *FDA Consumer*, December, 1980-January, 1981.
20. Morti Asner, "Worms That Turn Good Food into Bad," *FDA Consumer*, September, 1982.
21. F. Leo Kauffman, op. cit.
22. *Newsweek*, August 14, 1978, p. 8.
23. *Newsweek*, October 1, 1979, p. 24.
24. N. Irving Sax, *Dangerous Properties of Industrial Materials*, 2nd ed., Reinhold Publishing Corporation, New York, 1963.
25. *Chemical Week*, October 18, 1978, p. 33.
26. *Chemical Week*, January 10, 1979, p. 40.
27. *Food Chemical News*, January 15, 1979, p. 33.
28. Chris Lecos, "On Guard Against Radioactive Food," *FDA Consumer*, December 1980-January, 1981, p. 19.
29. John Tagliabue, "A Nuclear Taint in Milk Sets Off German Dispute," *The New York Times*, January 31, 1987, pp. 1, 4.
30. Jack R. Matches, Carlos Abeyta, "Indicator Organisms in Fish and Shellfish," *Food Technology*, June, 1983, p. 115.
31. Earl F. McFarren, "Assay and Control of Marine Biotoxins," *Food Technology*, March, 1971, p. 46.
32. S. A. Goldblith, "Pasteur & Truth in Labeling, 'Pro Bono Publico' — In the Best of Scientific Tradition," *Food Technology*, March, 1971, pp. 32, 33.
33. *Food Chemical News*, February 7, 1977, p. 38.
34. *Dairy Record*, May, 1982, p. 26.
35. Richard D. McCormick, "U.S. Must Be Allowed to Catch Up with Foreign Lead in Irradiation," *Food Development*, June, 1981, pp. 7, 8.
36. Michael G. Simic and Elliot DeGraff, "Radiation Chemistry Principles for Commercial Food Applications," *Food Development*, November, 1981, pp. 54, 55, 56, 58, 61, 66.
37. R. Paul Elliott, "Microbiological Control in Food Processing," *Food Product Development*, September, 1974, p. 48.
38. Karl Robe, "Bacterial Standards for Foods - Do They Work?" *Food Processing*, October, 1975, pp. 22-24.
39. Ari Brynjolfsson, Eugen Wierbicki, Cal Andres, "Irradiation Update," *Food Processing*, May, 1977.
40. Eugen Wierbicki, "Shelf Stable Irradiated Meat, Research on Wholesomeness Lays Groundwork for Future Development," *Food Development*, January, 1982, pp. 24, 25, 28.
41. Adel A. Kader, "Potential Applications of Ionizing Radiation in Post Harvest Handling of Fresh Fruits and Vegetables," *Food Technology*, June, 1986, pp. 117-121.
42. Rosetta L. Newsome, "Perspective on Food Irradiation," *Food Technology*, February, 1987, pp. 100-101.

43. J. E. Curtis and G. E. Huskey, "HACCP Analysis in Quality Assurance," *Food Product Development*, April, 1974, pp. 19, 24, 25.
44. Mario P. de Figueiredo, "Quality Assurance," *American Journal of Public Health*, May, 1971, p. 1019.
45. American Public Health Association, *Proceedings of the 1971 . . .*, op. cit.
46. *Life*, September 10, 1971, p. 36.

CHAPTER THREE
PRODUCT SPECIFICATIONS

Product specifications are like a map: they tell the processor what he is to make — his destination - and how best to achieve the desired results — the preferred route. These specifications dictate the required raw materials, processing aids, and manufacturing conditions as well as finished product attributes. Because these criteria are comprehensive, i.e., covering all phases of production, they are sometimes referred to as manufacturing specifications or "manuspecs" for short. The preparation of product specifications is an essential step in product planning.

Specifications for a product cannot be determined without a thorough familiarity with the food regulations. Products must meet the pertinent requirements for food additives, grades, and standards of identity. If these foods are designed for export they should comply with appropriate international codes, not to mention the regulations of the countries where they will be consumed. Although the food laws permit considerable flexibility and creativity in formulating new products, there are stringent directives which must be followed. Some background knowledge of these regulations is helpful, if not necessary, to provide guidance in product development.

PROCEDURE

By adhering to an orderly procedure for preparing product specifications, costly mistakes can be avoided. The manufacturing plant is in the best position to take a lead role in the preparation of these specifications. In the case of a new product and/or new process, the Research Department should provide the required experimental data, while the Quality Assurance Department should be responsible for gaining the necessary regulatory approvals. From the beginning of a new venture, the Marketing Department needs to be kept apprised of all decisions. To avoid any misunderstanding or recriminations at a later date, all parties must give their formal consent to the final plans. The following guidelines are suggested as a sound approach for developing product specifications.

Guidelines for Preparing Specifications

1. A HACCP review (see Chapter Two) should be made for all processes. If properly executed it will determine any hazard that may prevent the production of a safe and acceptable product, and it will identify the control points which are vital to reducing those hazards.

2. Grade, specifications, analytical methods for all raw materials, additives, processing aids and other supplies must be determined. Sampling plans including frequency of testing must be established. Approved sources of all materials must be located.

3. As an aid in following the manufacturing steps, a description of the process should be prepared. A flow diagram should be drawn in sufficient detail to indicate all control points, signifying those which are critical. For each control point, the plant should determine the test, frequency, specifications, method of analysis, and action to be taken.

4. Under finished product control, all specifications for the product including its labeling and packaging must be provided. (Labeling and packaging are discussed in subsequent chapters.) These specifications must be demonstrable under actual plant conditions. Grade, e.g., Food Chemicals Codex, batch or lot size, sampling plans, tests, analytical values and methods of analysis must be specified. In addition, typical analyses, bulk density, shelf life, corrosiveness, any dangerous properties, and other useful data need to be determined. Nutritional information may be required as well as data on applications and recommended use.

5. The regulatory status of all food ingredients shall be evaluated. New additives require approval from the Food and Drug Administration (FDA). If these products also are to be used in non-food applications, Premanufacture Notification may be required under the Toxic Substances Control Act. If needed for marketing, an application should be submitted for Kosher certification.

6. A compendium of analytical methods should be prepared. Tests included in this manual should be cross-referenced to raw material, in-process, and finished product specifications.

7. For ease of reference, proper records must be maintained by the plant. This information includes a complete set of product specifications, manufacturing instructions, formulations, and labels. FDA guaranties or certificates of compliance should be obtained from the suppliers of all raw materials and processing aids. As specified by Good Manufacturing Practice, laboratory test results, production records, and a sample of product from each lot should be retained for a period of time that exceeds the shelf life of the product except that they need not be retained for more than two years. Product safety information should be readily available to employees for all hazardous materials used in the plant.

8. Product specifications should be updated for each change, however slight, in the process. These modifications should be documented in all company records.

An Illustration for a Milk Replacer

The above procedure is best illustrated by again referring to the process for producing a milk replacer, first discussed in Chapter One. Figure 3.1 is a chart which has been prepared for one of the raw materials, sweet dry whey. For each test, the table shows the method of analysis, frequency of testing, specification, and action to be taken. Test values are prescribed by the U.S. Department of Agriculture (USDA) for U.S. Extra Grade, dry sweet whey, and by FDA for dry whey under the GRAS affirmation. Only approved methods of analysis can be used for these tests. In a similar manner, specifications can be documented for the other raw materials, the control points in the process, and the finished product, namely, the milk replacer.

Laboratory reports will be prepared on the basis of the given product specifications. In addition, a production record should be filled in for each batch of product as shown in Figure 3.2. These data will tie together the lot numbers and supplier for each raw material with the batch number of the finished product. Thus, if a complaint is received from a customer concerning a given batch of milk replacer, this problem can be traced back to the lot numbers of the ingredients that were used to blend the product. Laboratory test results can then be checked, and if necessary, the proper supplier can be notified of any difficulty.

Raw Material: Sweet Dry Whey   Specification No.: 1003

Grade: USDA Extra Grade   Effective Date: Jan 15, 1984   Approved by _JES_

Test	Method	Frequency	Specificiation	Action
Milkfat	AOAC 16.199	Spot check	NMT 1.5%	Retest and reject if out of spec.
Moisture	AOAC16.192	Each lot	NMT 5.0%	"
Std. plate count	USDA 918-109-2	Spot check	NMT 50,000/gm	"
Coliform	BAM ch. 5	"	NMT 10/gm	"
Scorched particles	USDA 918-109-2	"	NMT 15 mg	"
Titratable acidity	AOAC 16.023	Each lot	NMT 0.16%	"
Protein (N X 6.38)	AOAC 16.193	Spot check	NLT 11%	"
Physical appearance	Visual check	Each lot	Creamy white color, free from hard lumps	"
Alkalinity of ash	USDA 918-109-3	Spot check	NMT 225 ml	"
Ash	AOAC 16.196	"	NMT 12.5%	"
Heavy metals	FCC 3rd ed., pp. 512-513	"	NMT 10 ppm	"
Salmonella	BAM ch. 6	Every quarter	Negative	"

FIGURE 3.1. PRODUCT SPECIFICATIONS FOR A MILK REPLACER

Date	Batch Number	Quantity	Dry Whey			Soy Flour			Calcium Carbonate		
			Supply	Lot	Quantity	Supply	Lot	Quantity	Supply	Lot	Quantity

FIGURE 3.2. PRODUCTION RECORD FOR MILK REPLACER

PRODUCT SAMPLING

A valid testing protocol requires that the lot size, sampling plan, attributes, and methods of analysis all be determined. Each of these factors is interdependent, and therefore they should be considered together. The lot size will affect the frequency of sampling and in turn the number of analyses performed. Or working backwards, the cost and complexity of an analysis will influence the thinking about the sampling plan. Some optimum mix of these variables needs to be found so that maximum confidence can be achieved at a minimum required expenditure.

Determination of Lot Size

The definition for lot is given in the publication, *MIL-STD-105D*, Section 5.2 as follows:

> Each lot or batch shall, as far as is practicable, consist of units of product of a single type, grade, class, size, and composition, manufactured under essentially the same conditions, and at essentially the same time.[1]

When product is produced by a batch process, then quite obviously each lot will be identical to a batch. For a continuous process there is some leeway in specifying the lot size. This quantity may vary from 4,000 pounds for nonfat dry milk and other dairy products to as much material as is produced during an eight hour shift for less sensitive ingredients. The lot size for products historically free from pathogens might equal an entire day's production. The advantage of making the lot as big as possible is that the concomitant testing is reduced and shipping records are minimized. For much the same reasons customers prefer large sizes of lots.

On the other hand, there are definite drawbacks and some risks in increasing the lot size. Confidence is reduced, and in the event of a product recall, substantially greater amounts of materials must be taken back.

Sampling procedures all too often turn out to be the Achilles heel of quality assurance. Significantly different results can be obtained simply by modifying the method of sampling materials. This is the reason for the considerable discussion and not a little controversy surrounding this subject. The importance of good sampling has led to major efforts to improve techniques and to design better sampling apparatus.

Statistical Methods of Sampling

Because sampling relates to large numbers of data, statistical theory has been basic to its development. One of the earliest, widely used procedures was offered by the Department of Defense which published its authoritative bulletin *MIL-STD-105D*, updated April 29, 1963. This standard reference is relied upon by many government agencies and private parties. It permits the determination of values for Acceptable Quality Level (AQL) knowing the sample size and inspection level. A higher AQL or level of confidence can be obtained by taking more samples but not without increasing the associated analytical costs. Herein lies the need for developing inexpensive methods of testing. Chapter Eleven, for example, discusses the control of packaged weights using computerized check weighers. Other opportunities abound for increasing efficiency, e.g., by installing in-line processing controls, making use of rapid screening tests, and investing in automated laboratory instrumentation.

In cases where expensive and time consuming analytical procedures cannot be avoided and where resources are limited, a different sampling strategy may be employed. This approach depends on taking a single "representative" sample from each lot. Such a sample, by definition, is said to reflect exactly the composition of the universe or population. In theory this realization is possible since the lot was fixed as being uniform, but in practice unavoidable variations will exist within a lot, except possibly for gases. Even if the lot were initially homogeneous, some segregation of solids can be expected during materials handling through sifting and classification.[2] Recent studies have shown that larger particles in a heterogeneous mix will rise to the surface upon gentle shaking even though these particles may be denser and heavier. This phenomenon explains why Brazil nuts, for example, will rise to the top of a jar of mixed nuts.[3] Liquids also present problems as in special situations stratification can develop, sludges precipitate, or crystals form.[4]

Composite Samples

To get around these limitations of non-uniformity, elaborate procedures have been devised to obtain a composite sample which will closely approximate an ideal representative sample. Individual samples are withdrawn from different locations in the bulk container and blended together. This task can be performed manually, but to eliminate human bias a wide range of automatic sampling devices are being offered.[5,6] The point has been made, however, that an analysis of a composite sample will not reveal possible variations within the lot that could be significant. One reference notes that "Composite samples provide only limited information and the consequences should be carefully considered before deciding between this approach and the analysis of individual samples."[7]

Assuming that a processor has decided on using composite sampling and testing, everything will run smoothly as long as the results remain within the given specifications. But what does he do when an analysis falls outside the control limits? The instinctive reaction of many operators is to quickly grab another sample for retesting with the hope that the second report will be favorable, and reworking or other costly steps can be avoided. The fallacy of this reasoning has been exposed by showing that sooner or later, if enough samples are taken, a positive result will inevitably be obtained.[8] To avoid this pitfall an acceptable approach is to take two more samples and decide on the basis of what at least two out of the three analyses indicate.[9]

Certain tips on sampling may be helpful to the trainee and even to those of us with short memories. When removing a sample from a drum of solid material, do not skim product off the top as very likely a scoopful of material could have been added from another lot to make up the net weight. Take a thief sample instead from the top, middle, and bottom sections, or turn the drum on its side and roll it until the contents have been thoroughly mixed. Extreme care should be taken to ensure the integrity of any sample until it reaches the laboratory. Only clean apparatus and containers may be used, and the samples should be sealed until needed for analysis.

Frequency of Sampling

Economics ought to play an important part in developing sampling plans. Aside from questions of product safety, the functionality of a raw material is of primary concern. Particularly when an expensive ingredient is being used, the processor should know what attribute is critical to the performance of his finished product. He can then establish a schedule to measure this property. Thus, a blender of frozen dessert mixes will want to inspect closely incoming natural colloids for their stabilizing or thickening abilities in his formulations. A gel test, for example, will indicate the bloom of a gelatin or the milk protein reactivity of a carrageenan sample. By controlling this variable, not only will the manufacturer save money but he will have greater confidence in the quality of his ice cream base.

The frequency of sampling should be dictated above all else by reason and common sense. As pointed out in Chapter Two, scarce resources should be allocated to those control points that are critical to the process. As experience is gained with any operation, changes can be made in the sampling plan. When purchasing raw materials from a new vendor, initially every lot should be completely checked, but after five successive shipments have been found to be satisfactory, the frequency of testing may be reduced.[10] Then, only those analyses which are critical will be run for each lot, and other tests will be performed on a spot basis. (Figure 3.1 illustrates the testing for a milk replacer.) With increasing confidence in a supplier's dependability, this status may be formally recognized by accrediting him Vendor Certification. Experience has shown that while some suppliers are unreliable, others will ship product years on end without a single hitch.

TEST METHODS

Too often misunderstandings arise between parties because each is using a different analytical method. When standardized procedures are not used, care must be taken to correlate results with accepted methods. For most approved food additives, not only are the required tests specified, but the official methods of analysis are also given. These procedures have been selected only after they have been validated in practical applications and their methodologies have been thoroughly reviewed.

Origins of Analytical Methods

The scientific testing of foods dates back to the beginning of the nineteenth century. In 1800 a German immigrant, Fredrick Accum, established what was probably the first independent testing laboratory in England. He developed analyses for such food adulterants as alum, copper sulfate, and lead, and he compiled these methods in a book called *A Treatise on Adulterations of Food and Culinary Poisons*. In his zeal to expose the shady practices of the times he embarrassed not a few purveyors of food. Thus, when Mr. Accum was later indicted on minor charges of clipping pages from books in the library of the Royal Institute, there were mixed feelings about his disgrace. This episode led to the end of his career and his exile from England. Some time elapsed before Parliament passed in 1860 the first comprehensive food law, "An Act for Preventing the Adulteration of Articles of Food or Drink." Coinciding with revisions in this food legislation, The Society of Public Analysts was formed in 1875.

Meanwhile in the United States, W.O. Atwater, Professor of Chemistry at Wesleyan University in the 1870's, began his studies into the application of chemistry to food and nutrition. This interest led to his organizing the Connecticut Agricultural Experimental Station as well as similar stations in other states under the aegis of the federal government. He was appointed the first chief of the Office of Experimental Stations. During his career he collected extensive data on a wide variety of foods. This information was published under the authorship of Atwater and Bryant in 1899 and reprinted as USDA *Bulletin 28*. This reference work was revised and updated in 1950 as *Agriculture Handbook No. 8*, which is an authoritative source on the composition of raw, processed, and prepared foods.[11] This handbook is now available in serial form, beginning with No. 8-1, *Dairy and Egg Products*, issued in 1976, and extending through No. 8-16, *Legumes and Legume Products*, revised December, 1986.

Association of Official Analytical Chemists

Of equal significance was the organization of the Chemical Division in the Department of Agriculture in 1862. This group later became the Bureau of Chemistry and was put under the direction of Harvey W. Wiley. Recognizing the need for the standardization of analytical methods, Dr. Wiley together with several state chemists founded the Association of Official Agricultural Chemists in 1884. At that time there was considerable controversy between suppliers of recently introduced chemical fertilizers and the farmers who were buying these products. The new association helped to resolve disputes concerning the analyses of these new fertilizers, and thus began its long history of service to the agricultural and food industries. In 1965 the organization changed its name to the Association of Official Analytical Chemists, or AOAC as it is commonly known today.[12]

The AOAC collaborates with its members from government, private industry, and the universities to develop improved analytical procedures. Once these methods are approved, they are published initially in the association's *Journal*, and then in its *Official Methods of Analysis*, a compendium which is revised every five years and updated with annual supplements. In addition, AOAC puts out the *Bacteriological Analytical Manual (BAM)* which is a collection of qualitative and quantitative tests for microorganisms and certain of their metabolic products. These tests are particularly relevant to the quality assurance of foods.[13] The association performs all these functions without maintaining its own laboratory. For years it operated as a quasi branch of FDA, but an agreement signed in 1978 provided for the future conduct of its affairs to be at arms length from the government.

Food Chemicals Codex

Another pivotal organization is Food Chemicals Codex which was chartered in 1958 under the auspices of the National Academy of Sciences/National Research Council. Its avowed purpose is to provide a standard reference of additives for food processors, much like United States *Pharmacopeia* and the *National Formulary* serve the pharmaceutical industry. Before the founding of Food Chemicals Codex, each food manufacturer submitted its own detailed specifications when procuring ingredients. This practice had the disadvantage of not only being cumbersome but of lacking recognized standards of food safety. The first edition of the manual, *Food Chemicals Codex (F.C.C.)*, appeared in June, 1966, and was quickly accepted by FDA as the authoritative reference on ingredient specifications. The food regulations state in 21CFR170.30(h) that a food ingredient must comply "with any applicable food grade specifications of the *Food Chemicals Codex*." Thus "Food Grade" and "Food Chemicals Codex Grade" are synonymous in those instances where F.C.C. specifications exist.

(CFR stands for *Code of Federal Regulations* which is a compilation of effective regulations issued by agencies of the Executive Branch of the federal government. The CFR is composed of 50 titles. The title number is designated in front of the letters, CFR, when a reference is cited. Thus, Title 21 is reserved for food and drugs. The CFR is revised annually by the National Archives and Records Service of the General Services Administration in Washington, D.C. Before a final regulation is printed in CFR it is first published as a proposal along with explanatory material and comments in the *Federal Register*. The latter publication is issued daily and covers a wide range of public notices by the federal government.)

The third edition of *F.C.C.*, published in 1981, covers a total of 800 monographs on individual substances. The "General Provisions" of the manual should not be overlooked when referring to a single monograph. Thus a general requirement reads:

> Soluble substances, when brought into solution, may show slight physical impurities, such as fragments of filter paper, fibers, and dust particles, unless excluded by definite tests or other requirements; however, significant amounts of black specks, metallic chips, glass fragments, or other insoluble matter are not permitted.[14]

This specification can be interpreted as requiring that a pad test using filter discs be run on each such substance along with the other analyses.

Technical Grade Materials

On occasion the food chemist has recourse to technical grade materials. These products may be co-produced with food grade additives, and under controlled conditions they may be used in food processing. The American Society for Testing and Materials (ASTM) in 1980 initiated efforts to establish specifications for industrial grade chemicals, which until now have been sold under federal and military standards or individual customer specifications.[15] Except where otherwise mandated a food processor is free to use technical grade chemicals, but the burden of proof is placed on him to show that such materials are fit for processing into human food as stipulated in 21CFR110.80(a). To comply with this provision he would have to demonstrate that:

1. Any differences between the technical grade chemical and a comparable food grade chemical could be detected, and
2. Such differences could not be carried over into the final food product.

Not infrequently the only difference between a food grade additive and a technical product is their labels. A food grade chemical must include directions for its safe use and disclose any regulatory restrictions on its label.[16]

Other Analytical References

In addition to the references given above, there are a number of other useful guides available to the food scientist. *Compendium of Methods for the Micro-biological Examination of Foods*, edited by Marvin L. Speck, was prepared under contract with FDA. It reviews methods for determining food-borne infections and intoxications as well as food spoilage organisms.[17] *Standard Methods for the Examination of Dairy Products*, in its 14th edition, has been adopted by several states as the sole source of methods approved for compliance with FDA guidelines.[18]

The field of analytical chemistry is changing so fast that anyone responsible for this function needs to keep abreast of the latest developments. Rapid methods for microbiological analysis are receiving top interest as a means of streamlining lengthy plating techniques. Luminescent reactions for counting bacteria have been investigated with good results.[19] Another principle used is the measurement of radioactive carbon dioxide given off by bacteria.[20] Microcalorimetry can detect the minute amount of heat generated by metabolizing microorganisms.[21] These and other methods were reviewed during a symposium held by the Institute of Food Technologists in 1978.[22] To aid the microbiologist, kits are available for the rapid screening of certain pathogens and indicator organisms.[23]

SENSORY EVALUATION

The Pepsi Challenge, a promotional campaign which was run by PepsiCo, Inc. to advance its sales of cola beverage, did more than anything else to popularize sensory evaluation. Styled as the "most famous taste test in history," this campaign was introduced in 1974 in Dallas, Texas, and brashly run for over eight years nationwide.[24] Consumers across the country were blindfolded, handed two paper cups, one containing Pepsi-Cola and the other filled with its arch rival Coca-Cola, and then were asked to state their preference. The object, of course, was to publicize a strong liking for Pepsi and to get others to switch to that product.

With competition between colas never being static, The Coca-Cola Company responded in early 1985 by introducing a reformulation, promptly dubbed New Coke, of its classic soft drink. Having advance knowledge of this introduction, PepsiCo went through some agonizing moments until it could lay its hands on the new product. The first morning that New Coke could be bought in Atlanta, Georgia, six-packs were rushed by jet to PepsiCo's chairman, Donald Kendall, in Purchase, New York. Roger Enrico, the new president and CEO of PepsiCo, later recounted in his bestselling book the ensuing events:

"That afternoon, Pepsi employees — senior managers and secretaries, mailboys and executives — begin to gravitate toward Don Kendall's office. The cans of New Coke are opened. Paper cups are filled.

"All this happens in total silence. As everyone here is well aware, our future prosperity resides in those paper cups.

"Nervously, we all pause, cups in hand.

"Don Kendall takes the first sip.

"While he does, we sniff those paper cups as if they held prewar Chateau Lafite.

"Then we sip, letting the flavor expand on our palates.

"Seconds go by.

"We savor the carbonation.

"Only then do we swallow, reach for cups filled with water to refresh our taste buds, and start again."[25]

The Function of Taste Panels

Few taste panels convene in the chairman's office, but irrespective of the settings, panels routinely perform a vital function in the food industry. Notwithstanding advances in instrumentation - gas chromatography and mass spectrometry — only a human judge can integrate the perceptions of the five senses into a total sensory experience and in effect assume the role of an instrument. It is accepted that flavor is the sum total provided by taste, smell, sight, touch, and hearing.[26] Through the latter sense a person can appreciate products that are crispy or crunchy. Touch is necessary to distinguish texture: smooth or lumpy, hard or soft, chewy or crumbly. To perceive texture the panelist must move his jaws or mouth, just as he has to run his fingers over the surface of an object to get its feel. Color is important because of its associations — green with lime, red with strawberry/raspberry, and black with burnt.

Taste and smell, however, command the most attention from food sensory experts. As much as 75 to 80 percent of a flavor impression is smell, e.g., aroma, bouquet, or flavor notes.[27] Aroma is noticed not only in advance of eating but more importantly during chewing and swallowing. While humans do not have the same acuity for smell as dogs, for instance, our principal failing is the lack of a vivid vocabulary with which to describe odors. As a result we fall back on simile and metaphor.[28] For example, cheesy (like cheese) may be used to describe an aroma even though the product does not contain cheese.

Traditionally, four distinct tastes have been recognized: sweet, sour, salty, and bitter. Some academics, however, challenge the notion that there are only four tastes, asserting that this point of view is an a priori assumption that is self-serving. They would, for example, include the taste of monosodium glutamate, metallic taste, and still others. Furthermore there is controversy over whether tastes are synthetic, i.e. blendable like colors, which form new colors on mixing (purple from red and blue) or analytic like sounds that keep their individual identities when superimposed (drums and violins). The answer to this question will dictate the number of concepts — tastes and aromas — that must be identified and quantified to describe fully a given food product. Those concepts which are selected should be orthogonal, i.e., independent from each other.[29] For instance, if sweet and sour are included in the taste profile, then tart, which correlates positively with sour and negatively with sweet, would be redundant.

Some products have distinctive after-tastes, which will linger after the panelist has swallowed the food or spit it out. Mint or chili pepper will leave a cool/hot sensation in the mouth. Other products like lemon are astringent, and some, such as saccharin, contribute to a definite metallic taste. A notable example of a non-food product with strong after-taste is mouth wash. Because of its residual sensation, considerable time must elapse between sampling of this product.

For ages, flavorists, brewmasters, winemakers, and coffee tasters have held sway over their respective businesses.[30] While the principle of using a single technical expert is still valid, such persons are yielding to panels of trained professionals, typically six to ten in number, that can render a consensus decision. More credence is given by management to such verdicts.

Before an individual is accepted on a panel, he must familiarize himself with the rubric. Because the mind adapts to stimuli over a period of time, a zero drift in sensation occurs. Thus, as a panelist continues to sample a product, his perception of the flavor will diminish. In order to counter this effect, frequent rinsing of the mouth with water is required between tasting. To render qualified judgements, the panelist must acquire a thorough knowledge of the flavor concepts that are relevant to the products which he is tasting. By repeatedly sampling different reference substances that approximate each concept, he will get to know the meaning of a concept. Training of a skilled panelist usually requires many months.

Descriptive Analysis

The most sophisticated method of sensory evaluation is descriptive analysis. The profile of a product's flavor is developed by the following four steps.

1. All of the concepts — aromas, tastes, after-tastes — are identified for the product.
2. The intensity of each concept is determined.
3. The orders of appearance of the concepts are observed.
4. An overall impression or "amplitude" of the concepts is given.[31]

The above steps seem straightforward enough, but in reality they are extremely difficult to execute.

By virtue of their acceptance on the taste panel, members are equipped to sort out the concepts which encompass a food product. When it comes, however, to determining the intensity of each concept, the panel is literally confronted with the age old riddle of comparing apples and oranges. One approach, by ignoring such dissimilarities, looks at each concept independently. For example, each concept may be rated to be either not present, at its threshold perception, slight, moderate, or strong. Alternatively, an attempt can be made to integrate the data on intensity. The perception of a concept is compared to a universal scale of stimuli that is anchored at various points by known foods, e.g., grape juice, apple sauce, cinnamon gum.[32]

There are many ways of presenting data on intensities. Numerical values may be assigned to intensities by any one of several scaling procedures. The resulting values can be presented in tabular or graphic form. Perhaps the most eye-catching presentation is the Spidergraph®, a modification of which is illustrated in Figure 3.3.[33] Each spoke of the wheel represents a concept whose intensity is proportional to the length of the spoke. The ends of the spokes are connected to each other. Such a graph has been criticized, from a scientific point of view, for connecting unrelated points; however, this visual aid has been accepted as a useful tool.

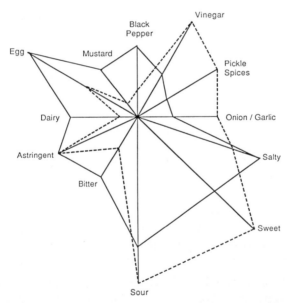

FIGURE 3.3. MODIFIED SPIDERGRAPH® FOR TWO MAYONNAISE SAMPLES, USING A UNIVERSAL SCALE TO DETERMINE INTENSITIES

The order of appearance of the different flavor concepts will depend not only on their properties but the manner in which they are released. In a cola beverage, for example, all flavor notes will be perceived almost simultaneously since the sample is swallowed in one gulp. On the other hand, a fatty food which melts and dissolves in the saliva may release certain of its flavors in a delayed sequence. Snack foods can be coated on their surfaces with salt to provide an immediate salty perception even though the average amount of salt in the product may be relatively low. Likewise, dried breakfast cereals can be sugar coated. Data on the order of appearance of flavor concepts can be presented on time-intensity charts to assist the food technologist in product formulation.[34]

The amplitude of a food product is judged on the basis of the presence of appropriate flavor notes and on the absence of off-notes. Equally important is the balance between these concepts and how well they blend together to give a unified impression. A product with a good amplitude will be more difficult to profile since the panelist will have more trouble in sorting out the different concepts. In spite of the problems and cost associated with descriptive analysis, this method has been found to be invaluable in new product development and in other applications where large differences between samples exist.

Difference Tests

Where variations in product are slight, difference tests are the methods used in sensory evaluation. Therefore, they are applicable in quality control in order to hold output within narrow limits. The following tests are the ones which are most discussed.

• **Paired Comparison Test,** as its name implies, involves two samples. The panelist is presented with the two samples and is asked to indicate which sample has a greater or lesser degree of a given concept, e.g., sweetness.

• **Triangle Test** is based on three samples, two of which are identical and the other differs in either a particular concept or in its overall flavor. In this test the panelist is requested to select the two identical samples leaving the "odd man out." The advantage of this test over the Paired Comparison Test is that there is no need to specify a concept.

• **Duo-Trio Test** again makes use of three samples of a product, two of which are the same and one is different. Either of the two identical samples is marked the reference or control sample. The panelist must choose which of the remaining two samples differs from (or is the same as) the reference.

• **Multiple Sample Tests** are those conducted with any number of samples. For example, a randomized mix of four control samples and four different ones can be given to the panelist who is then asked to sort them by their difference.[35]

In each of the above tests the panelist is presented with a forced choice. That is, in the Paired Comparison Test he would never be asked, "Is there a difference in sweetness?" Instead of this question, he would be queried, "Which is sweeter, sample 827 or 394?" This yes-no format eliminates the panelist's criteria for differences, and it generates data that can be treated by statistical methods. Only through statistics is an element of objectivity introduced into sensory evaluation.

The ability to detect differences are predicted by Weber's Law, which is expressed as follows:

$$jnd = \frac{\Delta R}{R}$$

where jnd is the Just-Noticed Difference, R is the magnitude of the stimulus and ΔR is the difference in the stimulus. To illustrate this relationship in a non-food system, the subtraction of one candle from ten lit candles would have the same effect as taking ten candles away from one hundred. The same relationship can be visualized in a food product where the number of chocolate chips in cookies effects the magnitude of the perceived chocolate flavor.[36] Results also depend on the sensitivities of the Difference Tests. The Paired Comparison is the most sensitive while the Multiple Sample Test with the highest number of samples is least sensitive. This effect is ascribed to "taste fatigue" whereby the mind has greater difficulty in keeping track of the more numerous samples.

Applications in Quality Control

Quality control depends on setting a value for a production target (reference or control) and setting limits that bracket this target. In establishing these limits, a manufacturer needs to determine consumer acceptance of production variations.[37] The latitude permitted by consumers quite logically would exceed the slight differences noted by panelists. In order to investigate sensory acceptance, experiments have been suggested whereby the sample is held fixed and the panel is replaced by numerous consumers. Thus, consumers become the variable in this experiment whereas samples were the variable in the procedures described previously.[38] In this manner confidence levels might be set for control charts. (A discussion of control charts is included in Chapter Eleven.). In the case of multivariant analysis, where more than one concept is followed at a time, the problem can best be visualized by a three dimensional model.[39]

Market research has some of the same outward appearances as sensory evaluation. In Paired Comparison tests, for example, consumers may be asked to state their preference for one product or another (as in the Pepsi Challenge.) Such hedonistic studies are conducted to elicit likes and dislikes. Market research departs from the principles of sensory evaluation because a consumer's choice is influenced by many external factors including price, packaging, promotion, prejudices, and social mores. Nevertheless, such studies can prove to be extremely valuable in product assessment as long as conclusions take into account the many variables.

GRADES AND STANDARDS OF IDENTITY

If we could turn the clock back to the 1920's, we would quickly gain an appreciation for the system of grades and standards which we now take for granted. Back then, the public was being victimized by an unscrupulous fringe of the food industry. These opportunists were selling cheapened products with inferior ingredients to unsuspecting consumers. One chronicler of the period reported that the fruit content of preserves went from 40 percent down to 30 to 25 or even 15 percent with the result that people were buying more and more water and pectin. In commenting on the scene, *Food Industries* magazine noted in 1929 that "the adulteration found today affects the pocketbook more often than the health."[40] Simultaneously wholesalers were frustrated in the conduct of their business by the lack of accepted grades for produce. They were having increasing difficulty in reconciling prices with the quality of goods received.

As attention was drawn to economic problems in this country, interest surged to protect consumers from gross deceptions and to establish a basis for fair dealings among distributors. Spearheading the drive for reform were those reputable food companies that were hurting from unfair competition. These legitimate businesses sought assistance from the federal government because the food industry was increasingly dominated by interstate commerce. The first tentative steps were taken when voluntary grades were adopted for potatoes in 1917, followed by butter in 1919, beef in 1926, and poultry in 1930. Since these grade standards were developed one by one the terminology lacked uniformity, a condition which survives in our regulations to this day.[41]

This system of voluntary standards was strengthened by the passage of important legislation. The McNary-Napes Amendment, known as the "Canners' Bill," was passed in 1930 to regulate substandard canned fruits and vegetables. Another milestone was the enactment in 1938 of the Federal Food, Drug, and Cosmetic Act which empowered FDA to promulgate Standards of Identity. Finally, the Agricultural Marketing Act of 1946 unified the supervision of the various grading procedures under the Department of Agriculture and the Department of Commerce.

Official Grades and Standards

As a consequence of the foregoing legislation we have the following structure of grades and standards:

Grade Standards, which are voluntary, have been issued for some 300 food and farm commodities by USDA. Examples of these grades are given in Table 3.1. USDA provides official grading services to packers and food processors for set fees. Products which have been so graded may carry the USDA grade shield or name, such as "U.S. Grade A." Even though these standards are voluntary, they have been widely accepted by the food industry.

Grade Standards for fishery products are controlled by the National Marine Fisheries Service of the U.S. Department of Commerce. Likewise, these grades are voluntary, and they may be indicated by labeling with the USDC grade shield.

Standards of Identity, which are mandatory, are enforced by FDA. These standards specify the compositions of a wide range of food products and establish

TABLE 3.1
USDA GRADE STANDARDS

Meat: Prime, Choice, Good*

Poultry: A, B, C

Eggs: AA, A, B

Butter, Cheese: AA, A, B

Nonfat Dry Milk: Extra

Fruits and Vegetables, fresh: Fancy, No. 1, No. 2

Fruits and Vegetables, Processed: Fancy or A, Choice or B, Standard or C

---

*A proposed name change would replace Good by Select. (*Food Chemical News*, March 9, 1987, p. 2.)

common names for them, such as mayonnaise, ketchup, and ice cream. Certain products, such as flour, may contain the word, "enriched," in their names, in which case they must contain the prescribed levels of added vitamins and minerals.

Standards of Quality are enforced by FDA for a number of canned fruits and vegetables. They limit, for instance, the "string" in green beans, peel in tomatos, hardness of peas, pits in pitted cherries, and "soupiness" in cream-style corn. These standards, which are mandatory, should not be confused with grades established by USDA.

Standards of Fill tell the packer how full a container should be to avoid deception and charges of slack filling. These mandatory standards are set by FDA.

Standards of Identity for meat products are controlled by USDA. These mandatory standards include corned beef hash and chopped ham.

Minimum Content Requirements have been established by USDA to specify the minimum percentage of a characterizing ingredient. Thus, beef stew must contain a minimum of 25 percent beef only. Otherwise the processor is free to create distinctive recipes.[42]

"Nutritional Quality Guidelines" and "Common or Usual Name"

The interest in establishing new food standards has subsided in recent years. One reason pertains to a general desire to reduce government regulation. Another cause is the use of alternative means of establishing specifications, for example, through the regulations for "Nutritional Quality Guidelines for Foods" (21CFR104) or for "Common or Usual Name for Nonstandardized Foods" (21CFR102). Under the latter provision there is a pending regulation for diluted fruit or vegetable juices and another proposal for vegetable protein products such as soy or peanut flour. In a recent affirmation of GRAS (which is discussed in a later section of this chapter) specifications were set for whey and modified whey products.

Safe and Suitable

In order to reduce the rigidity of recipe type food standards, a concept, known as Safe and Suitable, has been introduced. When included in a standard of identity, this provision allows the processor to choose from among many permitted functional ingredients. Thus, any qualified additive can be selected, the only conditions being that it is safe for the intended use and that it will perform an appropriate function, e.g., emulsification, leavening, or thickening. It may be used at a level no higher than necessary to achieve its intended purpose in the food. Because most of these additives are employed at less than 5 percent, this restriction was once considered but has since been abandoned. The Safe and Suitable provision is not meant to let a processor make a substitution that would degrade a product or change its character, for instance, by replacing the cream in ice cream with a vegetable oil. Rather, the flexibility provided by this concept is intended to encourage technological innovation for the ultimate benefit of consumers.[43]

Methods of Inspection

The determination of product grades has presented some special challenges in measurement. Since many of these grades were established on the basis of sensory or cosmetic differences, problems inevitably arose concerning subjective points of view. This need led to major programs to develop standardized procedures for grading. Although these efforts have largely been successful, this function still requires a high degree of human intervention. Now there is new instrumentation available which promises complete automation in grading.

Vision inspection, which uses optics and computers, can scan pizzas for burnt spots, crust height and shape, and the distribution of toppings. Or a candy bar manufacturer might check the integrity of its coatings for improved aesthetics as well as better product protection. Working on the same principle as candling eggs, an automatic vision inspection unit can grade vegetables. This method is also applicable to the trimming and sorting of food products. USDA jointly with a leading Midwestern university has been experimenting with grading and trimming beef. Vision inspection was able to show clearly the marbling and to delineate the fat from the red meat. In spite of the outstanding technical results, the sad commentary was made that "no action was taken beyond testing. A number of reasons were given why, including apprehension over a 'futuristic' technology."[44]

CODEX ALIMENTARIUS

International food standards are established by the Codex Alimentarius Commission for the guidance of its member nations, of which, at last count, there were 121 throughout the world. The commission was organized in 1962 by two bodies of the United Nations, the Food and Agriculture Organization (FAO) and the World Health Organization (WHO). It not only develops product standards but also codifies hygienic and technological practices, specifies methods of analysis and sampling, and establishes maximum limits for pesticide residues. In all, these guidelines cover processed, semiprocessed, and raw foodstuffs. The thrust of these regulations is twofold: they assist the less developed countries by supplying needed expertise, and they help to remove trade barriers between the industrialized nations.[45]

Adoption of Standards

A painstaking procedure has been adopted by Codex Alimentarius for developing its standards. A total of ten steps must be followed from the beginning when the commission decides there is a need for a standard to the last step of publishing the completed standard in the commission's organ. In between there are at least three review steps when drafts are submitted to the members for comments and actions. Because of these many safeguards, this process can take years for completion. In working on these projects the commission has fostered a new spirit of international cooperation, which is probably of even greater importance than the documents themselves.

Completed standards can be adopted by member nations in one of three ways. Full Acceptance means a country will allow the free movement of any food, of domestic origin or imported, which complies with the standard, and conversely, product which does not meet the standard will be excluded. Acceptance with Specified Deviations is when a country adopts a standard but stipulates additional requirements more stringent than the Codex conditions. Target Acceptance indicates that a nation intends to adopt a standard at a future date, but in the interim it will permit the unhindered exchange of goods meeting the standard. Although the goal of the commission is to obtain as many acceptances as possible, the mere existence of these prepared standards is of inestimable value.[46]

Acceptable Daily Intakes

One of the outstanding contributions the commission is making is in the area of food additives. Its Joint Expert Committee for Food Additives (JECFA) provides toxicological reviews and sets specifications for these additives. This committee defines Acceptable Daily Intakes (ADI) for food additives as a help in the proper use of these substances.[47] JECFA advocates a policy of "safety in numbers," namely, the wider the selection of foods consumed, the less chance there is that any one chemical will reach a hazardous level in the diet. This committee in 1967 stated:

> From the toxicological point of view there is less likelihood of long exposure, or of high or cumulative dose levels being attained, if a wide range of substances is available for use. Similar considerations apply to pesticides.[48]

FOOD ADDITIVES AND GRAS SUBSTANCES

The regulation of food additives has been a long, rocky road. Congress, ever taking a pragmatic approach, has been reluctant to act unless and until a patent need for reform has been demonstrated. The general regulation of food in the United States did not commence until 1906 when the Pure Food Act and the Meat Inspection Act were passed. In all probability neither of these landmark bills would have been adopted without the heroic efforts of three individuals. Dr. Wiley, head of the Bureau of Chemistry, organized his cohorts into a "Poison Squad" to disclose the most flagrant abuses of food adulteration. With equal force, Upton Sinclair in his muckraking novel, *The Jungle*, exposed the insanitary conditions which existed in the stockyards and meat packing plants of the Midwest. Finally, President Theodore Roosevelt, recognizing the urgency for action, threw the full prestige of his office behind the proposed legislation.[49]

The Pure Food Act

The Pure Food Act barred from interstate commerce any food that "contained any added poisonous or other added deleterious ingredient which may render such article injurious to health." This legislation had an immediate impact on the food industry, and considering the task at hand, it worked admirably well. It virtually put an end to further outbreaks of acute illnesses caused by the illegal addition of poison to food. This fact, however, was not accomplished without Herculean efforts. In each case of suspected adulteration, the burden of proof was placed on the government to show beyond doubt that the food in question was harmful. The penalties for infraction were relatively light so that the deterrents to repeated violations were minimal. Beyond these shortcomings, the law failed to regulate any natural toxins that might render a food dangerous, nor did it provide for the control of such unavoidable contaminants as pesticide residues.[50,51]

For those and other reasons the Pure Food Act was superseded in 1938 by the Federal Food, Drug and Cosmetic Act. Section 402 on adulteration was rewritten as follows:

> A food shall be deemed to be adulterated if it bears or contains any poisonous
> or deleterious substance which may render it injurious to health; but in case
> the substance is not an added substance such food shall not be considered
> adulterated under this clause if the quantity of such substance in such food
> does not ordinarily render it injurious to health. . . .

The two key words in the above definition are "may" and "ordinarily." Whereas an additive is prohibited if it *may* be harmful, however remote the possibility, (FDA interprets this language as requiring a safety factor in applying animal experimentation data to man of 100 to 1.[21 CFR 170.22]) all other substances are allowed if they *ordinarily* are not injurious to health. Thus, a bias was established against the use of all substances that are defined as food additives.[52]

In one other respect the food law passed in 1938 was glaringly deficient. Like the old Pure Food Act, the new legislation still required FDA to prove an additive was harmful before a food containing it could be removed from the marketplace. During the decades of the forties and fifties the task of enforcing this provision became much more difficult for two reasons. First, there was a vast increase in the number of additives that were being synthesized in food laboratories around the world, and second, health concerns shifted dramatically to such chronic diseases as cancer, circulatory ailments, and malnutrition. The burden of showing that an additive caused any one of these degenerative diseases became infinitely more complex than establishing a link with an acute illness. [53]

Food Additives Amendment of 1958

This weakness in the regulations was corrected by the Food Additives Amendment of 1958. In this watershed legislation the roles of government and industry were reversed. Instead of FDA having to prove an additive was dangerous, industry now was required to show that an additive was safe. [54] This requirement was spelled out in Section 409 as follows:

> A food additive shall, with respect to any particular use or intended use of such additive, be deemed to be unsafe . . . unless . . . there is in effect . . . a regulation issued under this section prescribing the conditions under which such additive may be safely used.

Not being content with this sweeping reform, Congress added, in an emotionally charged atmosphere, the further caveat that:

> No additive shall be deemed to be safe if it is found to induce cancer when ingested by man or animal, or if it is found, after tests which are appropriate for the evaluation of the safety of food additives, to induce cancer in man or animals.

This last change, now known as the Delaney Clause after Representative James J. Delaney who introduced it, has probably engendered more debate than any other provision in the food laws. Few will quarrel with the intent of the clause, but many question its approach. Much of the argument centers around an unproven hypothesis that even as little as one molecule of a carcinogen can trigger an irreversible reaction that will lead to cancer. Opposed to this view is the belief that there is a threshold level for any carcinogen, below which the substance has no effect whatsoever. Until more is known about the etiology of cancer, further debate on these irreconcilable positions seems futile. [55]

GRAS Substances

Appreciating the fact that the proposed amendment on food additives could cause severe dislocations in the food industry, Congress included a so-called grandfather clause exempting from regulation those foods with a long history of safe use. It states in Section 201:

> The term "food additive" means any substance the intended use of which results or may reasonably be expected to result, directly or indirectly, in its becoming a component or otherwise affecting the characteristic of any food (including any substance intended for use in producing, manufacturing, packing, processing, preparing, treating, packaging, transporting, or holding food; and including any source of radiation intended for any such use), if such substance is not generally recognized, among experts qualified by scientific training and experience to evaluate its safety, as having been adequately shown through scientific procedures (or, in the case of a substance used in food prior to January 1, 1958, through either scientific procedures or experience based on common use in food) to be safe under the conditions of its intended use.

The exclusion was the genesis of the Generally Recognized As Safe, or GRAS category of food substances. Also excluded, by other parts of Section 201, from the legal definition of a food additive are pesticides, new animal drugs, and color additives, each of these classes being regulated under other provisions of the act. In addition, substances which had received prior sanction of FDA or USDA before the Amendment was enacted were excluded. On the other hand, one should note that for the first time such indirect additives as used in packaging and processing, e.g., lubricants, (the regulations covering lubricants are given in 21CFR 178.3570) came under the statute, and radiation (or irradiation) was legally defined as an additive.

The response of FDA to this legislation was immediate. The agency, after publishing a proposed list of GRAS substances in the *Federal Register* and soliciting comments from "experts" about them, issued a final list of some 157 substances, which were later expanded by a second list of 27 more compounds. Simultaneously the Flavor and Extract Manufacturers' Association (FEMA) developed its own GRAS list of flavoring ingredients, which were acknowledged by FDA with few exceptions. In its haste to provide some guidelines for industry, FDA neglected the need to collect data for each substance on "the conditions of its intended use" as stipulated by the law. Also the agency was accused of being lax by not undertaking a systematic search of the literature for safety information. Both of these oversights later came back to haunt FDA.[56]

Color Additive Amendments

Not long after the Food Additives Amendment was passed, another issue came to a head. The United States Supreme Court, in a unanimous decision, upheld a directive by FDA to stop the continued use of a food color, namely, Red 32. The decision was based on the judgement that this color additive was shown to be harmful per se, notwithstanding the fact that under conditions of use it was completely safe. Quickly Congress realized that all certified food colors could be banned under this interpretation of the law, and it therefore passed new legislation in 1960, known as the Color Additive Amendments. (Section 706 of the amended Federal Food, Drug and Cosmetic Act) The new statute contained the wording of the Delaney Clause, but significantly the legislation provided for the provisional listing of color additives while they are under investigation.[57] By granting many extensions to several colors listed in this category, FDA has in effect granted quasi-approval of these colors.[58]

DES Proviso

Another conundrum, not unlike the ban placed on Red 32, resulted from the adoption of the Delaney Clause. Under the anti-cancer provision, diethylstilbestrol (DES), an undisputed carcinogen, would have been eliminated from all animal feeds. This compound, a synthetic form of the hormone, estrogen, had proven to be a remarkable growth promoter in raising cattle. Congress reacted in 1962 to this challenge by enacting the so-called "DES Proviso" which makes the anti-cancer clause inoperative for feed ingredients when:

(1) . . . such additive will not adversely affect the animals for which such feed is intended, and
(2) . . . no residue of the additive will be found (by methods of examination prescribed or approved by the Secretary . . .) in any edible portion of such animal after slaughter . . . . (409 [C] [3] [A])

DES was now spared, but the reprieve was only temporary. In 1971 an epidemiological study disclosed a high incidence of vaginal cancer in young women whose mothers had taken DES medication to prevent miscarriages. After renewed controversy and prolonged legal maneuverings, DES was banned for good in 1979.

Affirmation of GRAS List

In the meantime the unending dispute over food additives shifted to a new arena. In October, 1969, one of the substances included in the GRAS list, namely, the artificial sweetener, cyclamate, was shown in animal tests to be a weak carcinogen. An uproar immediately ensued, and even though GRAS substances are technically excluded from review under the Delaney Clause, FDA took hasty action to ban cyclamate. (A case history of artificial sweeteners is presented later in this chapter.) With the public's confidence in all food additives badly shaken, President Nixon, in his Consumer Message of October 30, 1969, ordered a full scale review of GRAS substances. A few weeks later in December, the White House Conference on Food, Nutrition and Health was convened to provide an open forum for airing concerns about food regulation.

Given its mandate to review the GRAS list, FDA set in motion the necessary machinery. It turned to the National Academy of Sciences to conduct an investigation of the usage of all GRAS substances. (The National Academy of Sciences, NAS, and its research arm, the National Research Council, NRC, fill the role of science referee on public issues. This august body was chartered by Congress in 1863 as a private, nonprofit organization although some 90 to 95 percent of its work has been done under contract for the federal government. NAS is involved with a number of food related topics, including Dietary Recommended Daily Allowance, ad hoc studies on nitrites and saccharin, and risk assessment.)[59] Next the agency contracted with private consulting firms to make literature searches for all safety data. This information was summarized in monographs, and these reports were submitted to a committee of the Federated American Societies for Experimental Biology (FASEB) for advisory opinions. Based on this input, FDA reaches its own binding decisions on the disposition of each substance. Those substances which are found to be acceptable are affirmed as GRAS in one of three categories:

(1) GRAS with no limitations other than good manufacturing practice where conditions of use do not differ significantly from those on which affirmation was based.

(2) GRAS with specific limitations in food categories, functional uses, and levels of use. Application beyond any of these limits would require a food additive petition.

(3) GRAS for a specific use without general evaluation of the use but subject to reconsideration upon such evaluation (i.e., spices, herbs).[60]

The meaning of "good manufacturing practice" for GRAS ingredients has been given by FDA as:

The requirements that a direct human food ingredient be of appropriate food grade; that it be prepared and handled as a food ingredient; and that the quantity of the ingredient added to the food does not exceed the amount reasonably required to accomplish the intended physical, nutritional, or other technical effect in food.[61]

Substances that do not meet the requirements for affirmation in one of the three GRAS categories or do not have a prior sanction will require a food additive regulation.

In undertaking the GRAS review, FDA categorically refused to be concerned with substances of natural biological origin such as fresh fruits and vegetables. Parenthetically it should be noted that only a limited number of GRAS substances out of an untold number of foods were itemized in the original GRAS lists. FDA felt that to attempt a blanket investigation under GRAS affirmation would be a gross misallocation of the nation's scientific resources. On the other hand, should safety questions arise or changes in processing or genetic makeup occur, such studies would be made.[62] With the increasing interest in genetic engineering, the issue of pre-market clearance for newly developed varieties or breeds is bound to assume greater importance.[63,64]

Cyclic Review of Food Additives

In 1980, after close to ten years of combined efforts, FASEB submitted its final report to FDA on GRAS substances and prior sanctioned ingredients. In all, 415 substances were evaluated.[65] While this massive undertaking was in progress, FDA launched a new program in 1977 to begin a "Cyclic Review" of all food additives. The fresh commitment was justified on the grounds that experimental data sooner or later become outdated and therefore must periodically be upgraded in the light of new scientific evidence.[66] Predating this review, a new class of additives, known as Interim Food Additives, had been created in 1972 for those substances with a history of use in food but about which questions of safety had been raised, and more information was sought. The GRAS review capped a record of outstanding accomplishments in food regulation; however, a thorny issue remains to be resolved as will be discussed in a subsequent section on Risk-Benefit Assessment.

UNAVOIDABLE CONTAMINANTS

Acknowledging that food will inevitably contain unwanted contaminants, the Federal Food, Drug and Cosmetic Act makes provisions for these imperfections. Under Sections 406 and 408, the Act allows for naturally occurring toxicants that are inherent constituents of foods, and it provides for the regulation of those unavoidable toxic substances that are added to foods either by the conscious acts of man, through environmental effects, or from other sources. Foods containing natural toxins may be consumed as long as the quantity of the poisonous material does not ordinarily render the food injurious to health. A well known example is the eating of potatoes which contain the toxic alkaloid, solanine.[67]

Tolerances and Action Levels

The food regulations make accommodations for the fact that with present technology, food cannot be grown and stored so that it is entirely free from impurities, e.g., pesticides. The standard of unavoidability is whether a given contaminant can be prevented by good manufacturing practice, to wit, proper control of all phases of production. For example, in 1978 FDA ruled that PCB's released in an industrial accident were avoidable and therefore not permitted in food.[68] To limit the levels of such impurities, either Tolerances or Action Levels are established. These specifications are determined on the basis of toxicity and unavoidability; however, FDA has declined to indicate what margin of safety it considers appropriate. Any food containing an impurity at or above a prescribed level will be considered adulterated and must be removed from the marketplace.

"Tolerances for Unavoidable Poisonous or Deleterious Substances" may be established under 21CFR109 through formal hearings. These actions are taken by the Environmental Protection Agency (EPA) when sufficient information is available concerning the toxicity of the impurity and when conditions contributing to the contamination are stable. In cases where data are incomplete or speed is of essence, e.g., during an emergency, FDA can announce Action Levels which take effect immediately without public hearings. Interested parties do have an opportunity to comment later, and these suggestions will be considered during a subsequent review. Whenever Tolerances or Action Levels are set, methods of analysis will always be specified.[69]

Natural or Unavoidable Defects

Similar proceedings can be taken under "Current Good Manufacturing Practice," 21CFR110, to establish limits on "Natural or Unavoidable Defects" in food. FDA has recently revised its procedure for setting a Defect Action Level (DAL) whereby a DAL will become effective for an interim period upon publication in the *Federal Register* or notification in the field. During the following year concerned persons are free to offer advice, and with sufficient reason these levels may be revised. DAL's are established taking into account the requirement that no hazards to health may be presented. Indicative of the type of defects covered, DAL's have been promulgated for histamine in tuna, insect fragments and rodent hairs in macaroni, mold in apricots, and thrips in sauerkraut.[70]

In the enforcement of Tolerances, Action Levels, or DAL's, FDA strictly forbids manufacturers from blending foods containing excessive amounts of a contaminant in order to bring them into compliance. This long standing policy has been knowingly disregarded only once, and that occasion was when acute shortages were likely to ensue. FDA has stated the rationale for this position as follows:

> The proper functioning of the agency's regulatory program requires that unlawfully contaminated food ordinarily be destroyed to make unlawful action unprofitable. The fact that the present owner of the food may not have been responsible for the avoidable contamination is not a convincing reason for departing from this policy.[71]

The negative impact of blending was amplified by former FDA Commissioner Donald Kennedy, who pointed out that "blending results in the redistribution rather than the elimination of a contaminant."[72]

Food manufacturers, faced with the unpleasant prospect of having to destroy nonconforming product, do have recourse to several alternatives. FDA has indicated that food, unsuited for human consumption, may nevertheless be fit for animal feed provided such a diversion does not pose a significant risk to animal or human health. Another possibility is to export the food on the condition that it conforms to all the laws and regulations of the country for which it is destined. Special permission might be received to sell contaminated food to a processor, such as a distiller, if there is no danger that the contaminant will be carried over to the finished products. Finally, the manufacturer should consider ways and means of reconditioning the food product so as to bring it within specifications.[73]

Action Levels have been established for a number of poisons, including aflatoxin, heavy metals like lead and mercury, and such persistent pesticides as aldrin, chlordane, and toxaphene.[74] Notwithstanding these measures taken, a report from the Office of Technology Assessment to Congress at the end of 1979 faulted FDA for not moving more aggressively to investigate potential contaminants.[75] This criticism has led FDA to establish a Surveillance Index (SI) for pesticides. Initially ten compounds have been selected as having uppermost priority, but this list will eventually be extended to monitor 300 plus agricultural chemicals.[76]

Two Controversial Decisions

Two controversial decisions illustrate the difficulty in applying the regulations on unavoidable contaminants. The first incident occurred on November 9, 1959, when FDA warned the public that a fraction of the cranberry crop was contaminated by the weedkiller, aminotriazole. This chemical was a properly registered pesticide, but no tolerance had been established for it in spite of formal requests by its manufacturer, American Cyanamid. The explanation given by FDA was that this compound was a reported carcinogen. Falling back on the wording of the Delaney Clause, Secretary of Health, Education and Welfare Arthur S. Flemming immediately invoked a ban of all lots of contaminated cranberries. With little additional information to go on, consumers were frightened into a nation-wide boycott of the entire cranberry crop. Besieged growers pleaded with the government for help, but assurances came too late from Washington to have an effect on Thanksgiving dinners — or Christmas either.[77]

A more recent case of contamination made front page news on February 3, 1984, when EPA ordered the suspension of further use of ethylene dibromide (EDB) on grain products. Long ago reported as a carcinogen in animals, this fumigant had been in use all this time on the erroneous assumption that there were no residues left in foods. In a bizarre ruling twenty years ago, FDA had exempted bromine compounds from regulation on the grounds that they were volatile and thus dissipated before foods reached the consumer. The fallacy of this reasoning was exposed by more sensitive analytical methods. Apparently undecided about the real threat posed by EDB, the Environmental Protection Agency, on announcing its ban, offered as guidelines to the states the following limits for EDB: 900 parts per billion (ppb) in raw grains, 150 ppb in flour and cake mixes, and 30 ppb in cereals, breads, and other ready-to-eat products.[78]

ARTIFICIAL SWEETENERS

The story of three sweeteners, cyclamate, saccharin, and aspartame, is more than just an amusing tale; it is a commentary on the effectiveness of our food regulations. Cyclamate was discovered accidentally in 1937 when Michael Sveda, a graduate student at the University of Illinois, noticed the sweetness of a compound he had in his laboratory. Abbott Laboratories introduced the sweetener in 1951, and it quickly became accepted by people on sugar-restricted diets. With the craze that followed for low calorie foods, the U.S. consumption of cyclamates (both the sodium and calcium salts were used) ballooned from 5 million pounds in 1963 to 15 million pounds in 1969. This growth, no doubt, would have continued unabated except for a dramatic turn of events.

Ban on Use of Cyclamates

In June, 1969, the results of an animal experiment at the University of Wisconsin were announced. Some mice had developed tumors when pellets containing cyclamate had been implanted in their bladders. Even though this study was considered to be inappropriate for testing oral toxicity, it immediately raised fears concerning the safety of cyclamate. Later that same year Abbott disclosed the results of a rat feeding study in which a mixture of cyclamate and saccharin in the ratio of 10:1, the same proportion as used in soft drinks, had been administered. Out of a population of 240 rats, eight which had been fed the highest dosage of sweeteners developed bladder tumors.

On October 18, 1969, Health, Education and Welfare Secretary Robert H. Finch announced the termination of the use of cyclamates in all food products, effective January 1, 1970, a date which was later extended. Invoking the Delaney Clause he pleaded that he had no choice in this matter. He could not consider such extenuating circumstances as the amount of the sweeteners fed in the rat test. At the highest level, the dosage was equivalent to a person consuming 350 bottles of diet soda daily. Further complicating the circumstances was the fact that two sweeteners of undetermined safety were given simultaneously to all of the rats in the feeding study. This experimental design left some doubt about which sweetener was responsible for the observed effects.

Convinced that the tests were invalid, Abbott proceeded to undertake an exhaustive investigation of cyclamate safety. It amassed sixteen volumes of data which it submitted in November, 1973, to FDA in support of a food additive petition. The agency responded by saying that more conclusive proof was required, but it gave no indication what would be acceptable evidence of safety. Abbott persisted in its request only to be asked by FDA in 1976 to withdraw its petition. Refusing to do so, the company requested a formal hearing which commenced in April, 1977. Eventually on September 4, 1980, FDA Commissioner Jere E. Goyan issued a final decree which denied Abbott's application. Admitting that safety is not defined by the law, FDA nevertheless found that the company's data failed to show that cyclamate does not cause cancer.[79] Resigned to this verdict, which culminated years of regulatory maneuvering, Abbott abandoned all further attempts to get approval for cyclamate.

Moratorium on Saccharin's Restriction

Turning back to saccharin, the other artificial sweetener that was tested with cyclamate, one finds that this food additive has been even more controversial.

Discovered in 1879, it ran afoul of Dr. Wiley's Bureau of Chemistry in 1912. The sweetener was banned from further use in foods although curiously allowed in chewing tobacco. Making a comeback, saccharin was reinstated as a food additive, and after the Food Additives Amendment was passed in 1958 it was included in the GRAS list. It escaped unscathed from the cyclamate fiasco in 1969, but on suspicion that it might have safety problems, saccharin was removed from the GRAS list in January, 1972.

Saccharin's prospects worsened when a study financed by the International Sugar Research Foundation revealed that several rats fed high doses of this sweetener developed bladder cancers. These results were substantiated by an FDA report in February, 1973. Both of these studies, however, were discredited because undetermined amounts of impurities were present in the saccharin samples. Tensions relaxed until March, 1977, when Canada alarmed the world by disclosing that saccharin caused cancer in rats during an extensive two generation feeding test using purified material. Immediately FDA announced its intention to prohibit the further use of saccharin except possibly as an over-the-counter drug.

Little did FDA suspect what the reaction of American consumers would be. The agency was inundated with some 100,000 letters mostly critical of its stand. Congress was pressured to take action to block a ban, and bowing to overwhelming public opinion it slapped an eighteen month moratorium on FDA's proposed restriction. Besides providing for a moratorium, the Saccharin Study and Labeling Act required warning labels on all foods containing saccharin and directed that a further investigation be made to resolve the issue. Little guidance was forthcoming as subsequent toxicological testing and epidemiological studies failed to refute the Canadian data or to indicate that saccharin presented a major health threat. As of 1987, Congress has been faced with the awkward task of extending the moratorium five times with no resolution of the issue in sight.[80]

The last chapter on saccharin has not yet been written. The emotion surrounding this additive might be better understood in light of people's dependence on an artificial sweetener. An estimated 50-70 million Americans consume low-calorie sweeteners with some frequency, hoping thereby to control their weight and improve their health. Moreover, an estimated 11 million diabetics have no choice but to restrict their sugar intake. In spite of saccharin's rather bitter aftertaste and its questionable safety, this sweetener will be in demand until a new and better substitute is available.

Approval of Aspartame

That potential replacement for saccharin is aspartame. The only requirement for finding a new sweetener appears to be serendipity — the faculty for making fortunate discoveries by accident. James M. Schlatter, a chemist at G.D. Searle, one day when licking his fingers, noticed the sweetness of a compound synthesized from two amino acids, aspartic acid and phenylalanine. Further studies could not have been more encouraging. Aspartame, as the sweetener is generically known, is almost 200 times as sweet as sugar, has a pleasant, clean taste, and is metabolized in the body like any protein. (Searle's tradenames for aspartame are Equal when it is used as a tabletop sweetener and NutraSweet when it is marketed as a food ingredient.) Only two drawbacks to the sweetener could be identified, its initial high cost and its instability at high temperatures or in strongly acidic solutions.

Armed with these generally favorable reports, in March, 1973, Searle applied to FDA for approval to market aspartame. The company anticipated an affirmative decision without delay. FDA did reply expeditiously, but no sooner was permission granted to use the sweetener than unexpectedly one lone scientist raised an objection. The agency felt compelled to stay its decision until all doubts could be resolved. The issue was whether aspartame would be harmful to persons suffering from phenylketonuria, a rare type of genetic disorder. The impasse over this contention lasted until 1981 when Searle initiated action to sue FDA. At this point a new commissioner took office, and upon reading the record he decided that the sweetener was indeed safe for most food products. This partial clearance for aspartame was extended in 1983 to include diet sodas, the major market for artificial sweeteners.

RISK-BENEFIT ASSESSMENT

The narrative about artificial sweeteners shows that under our food laws only those concerns which address risks can be weighed in making decisions. There can be no consideration given to the benefits that may be derived from using food additives. Not until consumers protested vociferously over FDA's proposal to ban saccharin did Congress listen and then only to pass temporary measures. When taken at face value, such skewed emphasis in our regulations can lead to absurdities. Only one person needs to demonstrate that he may be harmed by a food additive for that product to be summarily removed from the market. This one-sidedness in our laws ignores the fact that we are a pluralistic society with a broad continuum of wants and needs.

Constituents Policy, SOM, and De Minimis

The food industry has arrived at this unfortunate state of affairs somewhat unwittingly. In 1958 when the Delaney Clause was included in the Food Additives Amendment, the practical limit of detection for an impurity was 50 parts per million (ppm). By 1975, with improved analytical techniques, chemists were able to measure as little as 1 part per billion (ppb). This change was a fifty thousand fold increase in sensitivity, and the end of this progression is not yet in sight. Today the lower limit has been pushed to a few parts per trillion! Noting this trend in a speech given at the 1978 annual meeting of the Society of Plastics Industry, FDA Deputy Commissioner Sherwin Gardner opined, "We seem to be spending more and more time being concerned with less and less."[81] He went on to warn, "The level of acceptable risk should not be solely and arbitrarily a function of measurement technology."[82]

Struggling with these new realities, FDA is resorting to three artifices, the Constituents Policy, Sensitivity of Method (SOM), and *de minimis*. Each of these schemes attempts to define insignificant quantities of impurities. The Constituents Policy takes the position that contaminants found in a food additive can be ignored so long as the additive itself meets all specifications and criteria for safety. SOM, which was originally applied to animal drugs, defines a zero residue (or risk) as one which cannot be measured. The *de minimis* theory was introduced in a recent court case involving food packaging, and it argues that impurities present in trivial amounts can be disregarded.[83] As conceived, the Constituents Policy is the least restrictive of the three approaches, and therefore it has received the most attention as a legislative loophole.[84]

Risk Analysis

Risk assessment can best be illustrated by a graph. Figure 3.4 shows the risks for two food additives as being dependent on exposure. The basic proposition is that risk, regardless how it may be measured, is a function of the potency of

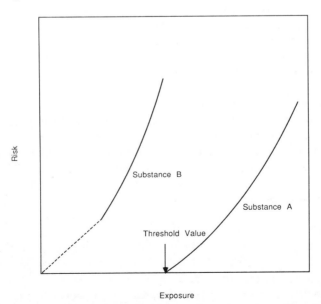

FIGURE 3.4. RISK ANALYSIS FOR FOOD ADDITIVES

the additive and the exposure to this compound. This relationship can be expressed by the following equation:

$$Risk = f \text{ (potency, exposure)}[85]$$

Referring again to Figure 3.4, one notes that Substance A possesses a threshold level of exposure below which no risk can be detected. On the other hand, Substance B is hypothesized to present a risk irrespective of how low the exposure may be. The solid part of the curve represents measured values of risk while the dotted line is a linear extrapolation to zero risk at no exposure.

Benefit Assessment

Because risks have dominated our thinking for so long, we have tended to overlook the truly remarkable accomplishments of food technology. We can dine on gourmet dishes at 41,000 feet or serve tastefully cooked meals at home in a matter of minutes using microwave ovens to heat up frozen entrees. Eating on the run at fast food restaurants or sampling ethnic cuisine in exotic surroundings are part of our lifestyle. Just as marvelous as these commonplace luxuries is the bounty which is shared by even the poorest members of society. Our populace, including infants, children, adults, and senior citizens, is the best fed of any generation.

None of these achievements would have been feasible without the use of food additives. Ironically, consumers have expressed their greatest distrust for those additives of greatest utility, namely, preservatives. Three classes of these preservatives stand out as being most controversial, antitoxidants (BHA/BHT), sulfites, and nitrites/nitrates. Looking first at BHA/BHT (abbreviations for butylated hydroxyanisole and butylated hydroxytoluene) these products are widely used antioxidants for vegetable oils and animal fats.[86] Added to such products as butter, margarine, and cooking oils, these additives retard rancidity. Without using antioxidants, food processors can resort to the partial hydrogenation of oils and fats, but this treatment which reduces the polyunsaturation is contrary to the advice of nutritionists. Statistics have drawn a parallel between the reduction of stomach cancer in certain countries and the increased ingestion of BHA/BHT antioxidants. Thus a 1982 report from Japan linking BHA to cancer was received with some skepticism.[87]

Sulfites, another class of preservatives, were used two thousand years ago by the Romans to produce wine. These compounds are selective antibacterial agents which will control mold but not inhibit the action of yeast. These functional additives are employed by food processors to prevent the browning of fruits and vegetables. Sulfites also possess a preservative effect on carotene and vitamin C.[88] Given these advantages, recent scattered reports about allergic reactions from sulfites are quite disturbing. Much of the blame is being placed on the misuse of sulfites in restaurant salad bars.[89] Concern about sulfites has led to a ban of these additives on fresh fruits and vegetables while permitting continued use of sulfites in most other applications.[90]

A third type of preservative, nitrates/nitrites, is used in curing such meat products as bacon, ham, and sausage. Not only do these compounds impart a red color and distinctive flavor to these products, but they provide unique protection against botulism.[91] For this reason there was considerable consternation several years ago when a report theorized that nitrates/nitrites, by forming nitrosamines, presented a cancer threat. Subsequent studies have dispelled much of this alarm, but the public's confidence is not yet restored. With similar frustrations in mind, John W. Hanley, former President and Chairman of Monsanto Company, told the Economic Club of Detroit in 1977:

> A responsible decision to permit chemical food additives must weigh any risks that remain after thorough testing against the benefits of protection against spoilage and bacteria, enhanced nutrition, greater convenience, lower cost, and on and on![92]

Risk-Benefit Accommodation

This concern about risk-benefit assessment in food regulation has been echoed by other responsible individuals. Former FDA Commissioner Alexander M. Schmidt observed, "In 1962, the first full benefit-to-risk accommodation was written into law with a requirement that pharmaceuticals be effective — that is to say 'beneficial' — as well as safe. . . . The same cannot be said for foods."[93] Nothing has changed in the intervening twenty-two years as indicated by the following excerpt from *Technology Review*.

> The regulation of risk as presently conducted by our federal government not only aggravates the hazards it's supposed to avoid, but stands as a major obstacle to technological innovation. New products are regulated much more strictly than older ones, impeding efforts to improve technology. . . . There is a politically feasible alternative: comparative risk regulation. Although most people think that new products and processes add to society's risk burden, in reality most new products don't "add to," they "substitute for." Yet regulatory agencies are prohibited from comparing the risks of a new product with the risks of the old ones for which it will substitute. . . . For example, cyclamates have been banned in this country while saccharin has not been. Canada has followed exactly the opposite course. One of us has banned the safer product and continues to use the more hazardous one.[94]

Referring to the same regulatory action over saccharin, Howard R. Roberts, former Acting Director, Bureau of Foods in FDA, noted, "The [proposed] saccharin ban was staunchly opposed by much of the public, and attention was focused more on the right of freedom of choice than on the right of freedom from risk." He went on to advocate the following procedure, based on risk-benefit assessment, to approve new food additives.

1. A petitioner would be responsible for submitting data on adverse effects together with an analysis of risk.
2 The same petitioner would be required to submit data on benefits by demonstrating the functionality of the additive at its intended level of use.
3. Opportunity for public participation would be assured through the current proposal/comment/final-order rulemaking process.[95]

Depending on the outcome of the above fact-finding procedure, the National Academy of Sciences outlined several options that would be available to FDA. The agency could ban the proposed additive altogether, accept it for restricted uses, limit its distribution to certain channels of commerce, require warning labels, or approve it unconditionally.[96]

The Reagan Administration has gone on record as lending full support to the reform of the Federal Food, Drug and Cosmetic Act. Former Secretary of Health and Human Services Richard S. Schweiker noted that the benefits of a food additive may justify its use notwithstanding some risks that may be posed by it. In this connection he observed:

> We have to recognize that the public neither wants nor expects the government to guaranty "zero risk." We have to recognize that absolute, 100 percent safety is impossible, in food as well as every other aspect of a consumer society.[97]

REFERENCES

1. Department of Defense, *Military Standard Sampling Procedures and Tables for Inspection by Attributes*, Bulletin *MIL-STD-105D*, Washington, D.C., April 29, 1963.
2. Jerry R. Johanson, "Particle Segregation . . . and What To Do About It," *Chemical Engineering*, May 8, 1978, pp. 183-188.
3. James Gleick, "Sometimes Heavier Objects Go to the Top: Here's Why," *The New York Times*, March 24, 1987, pp. C1, C11.
4. Morton E. Bader, "Quality Assurance — II, The Quality-Control Laboratory," *Chemical Engineering*, April 7, 1980.
5. F. Jordison, "Tips on Automatic Sampling," *Chemical Engineering/ Deskbook Issue*, October 30, 1978, pp. 103-107.
6. Gerald Fitzsimmons, Wm. S. Stinson, "Automatic Sampling Eliminates Guesswork," *Food Processing*, April, 1979, pp. 128, 130.
7. Byron Kratochvil, John K. Taylor, "Sampling for Chemical Analysis," *Chemtech.* September, 1982, pp. 564-570.
8. Hy Pitt, "The Resampling Syndrome," *Quality Progress*, April, 1978, pp. 27-29.
9. Stanley A. Marash, "Some Considerations for Bulk Sampling," *Food Product Development*, September, 1980, pp. 56, 57.

10. Morton E. Bader, op. cit.
11. Bernard L. Oser, "The Impact of Analytical Chemistry on Food Science," *Food Technology*, January, 1975, pp. 45-47.
12. Cedilia Cassidy, "Quality Assurance through Methods Standardization — The Role of AOAC," *Food Product Development*, June, 1978, pp. 24, 25, 28.
13. *ASM News*, Vol. *44*, No. 9, (1978), p. 479.
14. *Food Chemicals Codex*, 3rd ed., National Academy Press, Washington, D.C., 1981.
15. "It's Going to Be Easier to Sell to Uncle Sam," *Chemical Week*, January 21, 1981, pp. 44, 45.
16. "Food Grade Chemicals without Directions for Use May Be Misbranded," *Food Chemical News*, July 28, 1986, p. 15.
17. Marvin L. Speck, *Compendium of Methods for the Microbiological Examination of Foods*, American Public Health Association, Inc., Washington, D.C., 1976.
18. Elmer H. Marth, *Standard Methods for the Examination of Dairy Products*, 14th ed., American Public Health Association, Inc., Washington, D.C., 1978.
19. Katherine F. Baker, Richard D. McCormick, "Rapid Screening, Quantitation of Microorganisms by Bioluminescence," *Food Product Development*, June, 1976, pp. 93, 94.
20. "New Instrument Speeds Microbial Detection and Count," *Food Engineering*, May, 1976, p. 86/Int-62.
21. Paxton Cady, "Instrumentation in Food Microbiology," *Food Product Development*, April 1977, pp. 80, 82, 83-85.
22. "Rapid Methods and Automation for Food Microbiological Analysis," *Food Technology*, March, 1979, pp. 52-74.
23. Robert J. Swientek, "Rapid Bacteria Identification Kits Gaining Food Industry Acceptance," *Food Processing*, April, 1981, pp. 160-164.
24. Roger Enrico, *The Other Guy Blinked*, Bantam Books, New York, 1986, p. 86.
25. Roger Enrico, op cit., p. 217.
26. Camille E. Appel, "Taste ≠ Flavor," *Chemtech*, July, 1985, pp. 420-423.
27. Ibid.
28. Boyd Gibbons, "The Intimate Sense of Smell," *National Geographic*, September, 1986, pp. 324-361.
29. Gail V. Civille and Harry T. Lawless, "The Importance of Language in Describing Perceptions," a paper presented at the 46th Annual Meeting of the Institute of Food Technologists, in Dallas, Texas, June 18, 1986.
30. Bill Dyer, Winemaker, Sterling Vineyards, P.O. Box 365, Calistoga, CA 94515, private communication, November 26, 1986.
31. Jean F. Caul, "The Profile Method of Flavor Analysis," *Advances in Food Research, 7,* pp. 1-40, (1957).

32. Gail V. Civille, "Sensory Descriptive Flavor Analyses," Manual from a course given at The Center for Professional Advancement, East Brunswick, New Jersey, December 15-18, 1986, p. C10.
33. Herbert Stone et al., "Sensory Evaluation by Quantitative Descriptive Analysis," *Food Technology*, November, 1974, pp. 24-34.
34. William E. Lee III and Rose M. Pangborn, "Time-Intensity: The Temporal Aspects of Sensory Perception," *Food Technology*, November, 1986, pp. 71-78, 82.
35. Herbert Stone and Joel L. Sidel, *Sensory Evaluation Practices*, Academic Press, Inc., Orlando, Florida, 1985, pp. 136-143.
36. Howard R. Moskowitz, *Product Testing and Sensory Evaluation of Foods*, Food & Nutrition Press, Inc., Westport, CT, 1983, pp. 286-288.
37. J. L. Sidel, H. Stone, J. Bloomquist, "Industrial Approaches to Defining Quality," *Sensory Quality in Foods and Beverages*, edited by A. A. Williams and R. K. Atkin, Verlag Chemie International, Deerfield Beach, Florida, 1983, p. 53.
38. Michael A. P. O'Mahoney, "Sensory Evaluation in Food Product Development," manual for a course sponsored by Monterey Seminar Group, Carmel, California, November 14-15, 1986, pp. 6, 7.
39. J. J. Powers, V. N. M. Rao, "Computerization of the Quality Assurance Program," *Food Technology*, November, 1985, pp. 136-142.
40. "Regulate Us! Cried Food Companies in 1928," *Food Engineering*, October, 1978, pp. 97, 98, 100.
41. Mel Seligsohn, "Food Grading: Overhauling an Old System," *Food Engineering*, September, 1980, pp. 40-42.
42. U.S. Department of Agriculture, "Federal Food Standards," Bulletin *AMS-548*, November, 1972.
43. Howard R. Roberts, "FDA Status: Overview and Insight," paper presented at the 1977 International Association of Ice Cream Manufacturer and Milk Industry Foundation Convention, Denver, Colorado, October 25, 1977.
44. James N. Wagner, "Inspecting the 'Impossible'," *Food Engineering*, June, 1983, pp. 79-92.
45. Eddie F. Kimbrell, "Codex Alimentarius Food Standards and their Relevance to U.S. Standards," *Food Technology*, June, 1982, pp. 93-95.
46. David Garten, "Food Standards for the World Community Are Within Reach," *Food Product Development*, July, 1979, pp. 60, 61.
47. G. B. Woodin, "Codex Alimentarius Develops Global Quality Safety Specifications," *Food Development*, December, 1981, p. 24.
48. "Naturally Occurring Toxicants in Foods," *Food Technology*, March, 1975, pp. 67-72.
49. James Harvey Young, "The Long Struggle for the 1906 Law," *FDA Consumer*, June, 1981, pp. 12-16.
50. Roger D. Middlekauff, "200 Years of U.S. Food Laws: A Gordian Knot," *Food Technology*, June, 1976, pp. 48, 50, 52, 54.

51. Food Safety Council, "Principles and Processes for Making Food Safety Decisions," *Food Technology*, March, 1980, pp. 81-125.
52. Roger D. Middlekauff, "Legalities Concerning Food Additives," *Food Technology*, May, 1974, pp. 42, 44, 46, 48.
53. Virgil O. Wodicka, "Food Safety in 1973," *Food Product Development*, July-August, 1973, pp. 48, 50, 51, 52.
54. "How It All Began," *Food & Drug Packaging*, November, 1979, pp. 66, 67.
55. "Should the Delaney Clause Be Changed?" *Chemical & Engineering News*, June 27, 1977, pp. 24-46.
56. Richard L. Hall, "GRAS — Concept and Application," *Food Technology*, January, 1975, pp. 48, 50, 52, 53.
57. Harry N. Meggos, "Colors — Key Food Ingredients," *Food Technology*, January, 1984, pp. 70, 73, 74.
58. "Color Additive Provisional Listings Again Upheld by District Court," *Food Chemical News*, March 3, 1986, pp. 8, 9.
59. Donald E. Veraska, "How Useful Is Slow-Moving NAS?" *Chemical Business*, June 28, 1982, pp. 29-32.
60. Mary T. O'Brien, "FDA Finalizes GRAS Criteria, Affirms 12 GRAS Substances," *Food Product Development*, February, 1977, p. 10.
61. "Changes in GRAS Affirmation Regulations Proposed by FDA," *Food Engineering*, November, 1982, p. 15.
62. "FDA Issues 47-pg GRAS Status Report," *Food Processing*, November, 1974.
63. A. Kramer, "Changes in Food Quality, 1927-1977," *Journal of Food Quality*, Vol. *1*, No. 1 (1977), pp. 1-4.
64. Edward L. Korwek, "FDA Regulation of Food Ingredients Produced by Biotechnology," *Food Technology*, October, 1986, pp. 70, 72, 74.
65. *Food Chemical News*, May 26, 1980, pp. 48, 49.
66. "FDA's 'Cyclic Review' Launched This Month," *Food Engineering*, March, 1977, p. 22.
67. *Food Chemical News*, April 23, 1979, p. 4.
68. *Food Chemical News*, April 17, 1978, p. 10.
69. Alexander M. Schmidt, "How the Contamination of Food Is Controlled," *Dairy Industry News*, January 17, 1975, pp. 9, 10.
70. "New Procedure for Defect Action Levels," *Food Technology*, November, 1982, p. 40.
71. *Food Chemical News*, October 3, 1977, pp. 40-46.
72. *Food Chemical News*, February 6, 1978, pp. 3-5.
73. "Do World Food Needs Demand a Reconsideration of Policies on Blending?" *Food Product Development*, October, 1975, pp. 22-32.
74. Food and Drug Administration, *Action Levels for Poisonous or Deleterious Substances in Human Food and Animal Feed*, Washington, D.C., June, 1978.
75. "Food Contamination: OTA Urges Changes," *Chemical Marketing Reporter*, December 17, 1979, pp. 4, 57.

76. *Food Chemical News*, March 3, 1980, p. 15.
77. "Cranberry Smash," *Science*, Vol. *130*, No. 3387, November 27, 1959.
78. "U.S. Prohibits Use of Pesticide Tied to Animal Cancer," *The New York Times*, February 4, 1984, pp. 1, 9.
79. Chris Lecos, "The Sweet and Sour History of Saccharin, Cyclamate, Aspartame," *FDA Consumer*, September, 1981, pp. 8-11.
80. "FDA Reviewing Saccharin Research to Determine Moratorium Position," *Food Chemical News*, April 6, 1987, pp. 30, 31.
81. Marie Attmore, "Regs Should Assess Risk-Benefit Ratio, Says FDA Official," *Food and Drug Packaging*, May 18, 1978, pp. 1, 32.
82. "Gardner Sees Risk Assessment As Solution to Packaging Dilemma," *Food Chemical News*, April 17, 1978, p. 20.
83. Robert J. Kelsey, "New FDA Rule Re-Interprets Delaney Clause," *Food and Drug Packaging*, June, 1982, pp. 6, 29.
84. "Constituents Policy Would Affect Indirect Additives, Color Additives," *Food Chemical News*, April 5, 1982, pp. 3-9.
85. Ronald J. Lorentzen, "FDA Procedures for Carcinogenic Risk Assessment," *Food Technology*, October, 1984, pp. 108-111.
86. "Synthetic Anti-Oxidant Food Market Threatened by EEC Draft," *European Chemical News*, November 28, 1975, p. 32.
87. *FDA Consumer*, September, 1982, p. 22.
88. "Sulfites as Food Additives," *Food Technology*, October, 1975, pp. 117-120.
89. "A Growing Concern Over Sulfites," *Chemical Week*, November 9, 1983, p. 16.
90. Irvin Molotsky, "U.S. Issues Ban on Sulfites' Use in Certain Foods," *The New York Times*, July 9, 1986, pp. A1, A16.
91. "Additives on Way Out?" *Chemical Week*, October 29, 1975, p. 17.
92. John W. Hanley, "Has Emotion Tipped the Scales on Consumer Safety?" *Vital Speeches of the Day*, November 15, 1977, pp. 92-95.
93. Alexander M. Schmidt, "How FDA Views the Benefit/Risk Question," *Food Product Development*, February, 1975, pp. 78, 82, 85.
94. Peter Huber, "Discarding the Double Standard in Risk Regulation," *Technology Review*, January, 1984, pp. 10, 11, 14.
95. Howard R. Roberts, "Regulatory Aspects of the Food Additive Risk/Benefit Problem," *Food Technology*, August, 1978, pp. 59-61.
96. "For Saccharin Decision, Revamp Food Laws," *Science News*, March 10, 1979, p. 150.
97. "Schweiker Says Administration Will Help Develop Food Safety Bill," *Food Chemical News*, February 8, 1982, p. 23.

CHAPTER FOUR
LABELING

Food labeling refers to all written, printed, and graphic material displayed upon a food article, its wrapper, or container; or accompanying said article in commerce. The preparation, review, and approval of such labeling is the most contentious activity in quality assurance. The food manufacturer in his zeal to promote his product, understandably will make such claims on his labels and in advertising that he feels will attract the consumer's attention and loyalty. Although the preponderance of this promotion speaks to the legitimate concerns of the consumer, some copy blatantly exceeds accepted ethics by exploiting the public's ignorance and fears. The Institute of Food Technologists (IFT), the professional society for food scientists, decries such tactics and has singled out four types of claims as being particularly offensive:

- Claims which state or imply that "natural" foods are superior to processed foods.
- Claims which disparage "food additives."
- Claims which highlight unproven health benefits.
- Claims which make emphatic nutritional claims about foods with dubious nutrient content, or which unfairly compare nutritional benefits of different foods.[1]

As food manufacturers persist in probing the limits of what is permitted by regulation or propriety, they can expect to encounter further criticism from competitors as well as consumers.

Mirroring technological progress, swings in social mores, and political trends, labeling practice is in a constant state of flux. In order to address current labeling issues, the Food and Drug Administration (FDA), U.S. Department of Agriculture (USDA), and Federal Trade Commission (FTC) jointly held a series of public hearings in 1978. These forums were conducted in five different cities across the country with the aim of obtaining the widest participation: Wichita, Kansas; Little Rock, Arkansas; Washington, D.C.; San Francisco, California; and Boston, Massachusetts. The agenda covered seven controversial subjects:

1. Ingredient labeling
2. Nutritional labeling and other dietary information

3. Open date labeling
4. Imitation and substitute foods
5. Food fortification
6. The total food label
7. Safe and suitable ingredients.[2]

Based on the outcome of these hearings, FDA proposed a number of limited changes in the labeling regulations, but it was stymied in making any sweeping reforms because of restrictions in the food laws. Even these modest proposals, however, were largely dropped because of a lack of consensus. These issues nonetheless are still very much alive and form the basis for much of the discussion which follows, except that the concept of Safe and Suitable has already been reviewed in Chapter Three under Grades and Standards.

LEGISLATION

Section 403 of the Federal Food, Drug and Cosmetic Act (FD&C Act) states that a food product shall be deemed to be misbranded and therefore prohibited:

(a) If its labeling is false or misleading in any particular. . . .
(b) If it is offered for sale under the name of another food.
(c) If it is an imitation of another food, unless its label bears . . . the word "imitation". . . .
(d) If its container is so made, formed, or filled as to be misleading.
(e) If in package form unless it bears a label containing (1) the name and place of business of the manufacturer, packer, or distributor; and (2) an accurate statement of the quantity of the contents in terms of weight, measure, or numerical count. . . .
(f) If any word, statement, or other information required by . . . this Act to appear on the label . . . is not prominently placed thereon . . . .
(g) If it purports to be or is represented as a food for which a definition and standard of identity has been prescribed . . . unless (1) it conforms to such definition and standard, and (2) its label bears the name of the food specified in the definition and standard, and . . . the common names of optional ingredients (other than spices, flavoring, and coloring) present in such food.
(h) If it purports to be or is represented as — (1) a food for which a standard of quality has been prescribed . . ., and its quality falls below such standard . . .; or (2) a food for which a standard or standards of fill of container have been prescribed . . ., and it falls below the standard of fill
. . . .
(i) If it is not subject to the provisions of paragraph (g) of this section unless its label bears (1) the common or usual name of the food, if any there be, and (2) in case it is fabricated from two or more ingredients, the common or usual name of each such ingredient; except that spices, flavorings, and colorings . . . may be designated as spices, flavorings, and colorings without naming each . . . .

(j) If it purports to be or is represented for special dietary uses, unless its label bears such information concerning its vitamin, mineral, and other dietary properties . . . necessary in order fully to inform purchasers as to its value for such uses.

(k) If it bears or contains any artificial flavoring, artificial coloring, or chemical preservative, unless it bears labeling stating that fact . . . .

(l) If it is a raw agricultural commodity . . . containing a pesticide chemical applied after harvest, unless the shipping container of such commodity bears labeling which declares the presence of such chemical . . . .

(m) If it is a color additive, unless its packaging and labeling are in conformity with . . . Section 706.

(n) If its packaging or labeling is in violation . . . of the Poison Prevention Packaging Act of 1970.

(o) If it contains saccharin, unless . . . its label and labeling bear the following statement: "Use of this product may be hazardous to your health. This product contains saccharin which has been determined to cause cancer in laboratory animals.". . . .

(p) If it contains saccharin and is offered for sale, but not for immediate consumption, at a retail establishment, unless such retail establishment displays prominently . . . notice . . . required by paragraph (o) to be on food labels and labeling . . . .

Discretionary Authority of FDA

Beyond the powers over food labeling invested in FDA by reason of Section 403, the agency has assumed wide discretionary authority to promulgate regulations reasonably related to the FD&C Act's purpose. In taking these actions, FDA has cited Section 701 (a) which states:

The authority to promulgate regulations for the efficient enforcement of this act, except as otherwise provided in this section is hereby vested in the Secretary (FDA).

Thus, the agency has specified extensive rules for nutritional labeling, all without the express consent of Congress.[3]

Two important exceptions to the labeling requirements are spelled out in the FD&C Act. Section 405 exempts from the labeling regulations any food which in accordance with the practice of the trade is to be repacked, labeled, or processed in substantial quantities at another establishment. Section 801 says that product intended for export shall not be considered to be misbranded if it meets the specifications of the buyer, is not in conflict with the laws of the country to which it is to be exported, and is labeled for export.

Role of USDA in Labeling

Under the Federal Meat Inspection Act, the Poultry Products Inspection Act, and the Egg Products Inspection Act, USDA is given jurisdiction over products covered by these laws. The labeling requirements for these products parallel those specified by the FD&C Act with one notable difference. USDA interprets its authority under the legislation as requiring the prior approval of all labels by the department before their application in commerce. Under this system, a meat packer must submit to USDA for approval a sketch of a proposed label with the following information:

1. Product name
2. Ingredient statement and/or qualifying statement
3. Inspection legend
4. Firm name and address including zip code
5. Refrigerated or frozen statement (if necessary)
6. Net weight statement for products sold at retail intact.[4]

Fair Packaging and Labeling Act

Notwithstanding the encompassing labeling requirements of the FD&C Act, it was found wanting in protecting the public from certain deceptive practices. According to legend, U.S. Senator Philip A. Hart from Michigan one day counted the number of cherries in a frozen cherry pie which he had purchased at a supermarket. Somewhat to his amazement, he found that the quantity of cherries in no way compared with the number indicated on the label. This rather mundane observation led him to become engrossed in consumer protection. During his investigations he uncovered many dubious practices, for example, the indiscriminate use of the term "Super Family Economy Size" for a variety of actual package sizes and illustrations on food cartons which frequently did not reflect the true contents of the container. After six years of proselytizing, Senator Hart finally succeeded in mobilizing Congress to pass a major labeling bill in 1966.[5]

The Fair Packaging and Labeling Act (FPLA), which was championed by Senator Hart, has as its overriding purpose the facilitation of "value comparisons." Administration of the law is shared by three government agencies: FDA, which oversees all food, drug, and cosmetic products; FTC, which is responsible for other consumer products; and the Department of Commerce (DOC), which has been given the specific assignment of standardizing packaging sizes. Meat and poultry products are specifically exempt from the FPLA inasmuch as they are regulated under other statutes. The FPLA is the basis for several important regulations.

• The Principal Display Panel (PDP) is defined as that part of a label that is most likely to be displayed, presented, shown, or examined under normal and customary conditions of display for retail sale. Certain information, such as a statement of the identity of the commodity, is required to appear on the PDP. (21CFR101.1)

• Ingredients must be listed by common or usual name in descending order of predominance by weight. (21CFR101.4) Deviations from this rule apply to products containing an expensive ingredient (e.g., wild rice in a blend of brown rice), the percentage of which must be designated. (21CFR101.3)

- The name and place of business of the manufacturer, packer, or distributor must be conspicuously shown. The street address must be included unless the distributor's name is listed in the local telephone directory. If the name appearing on the label is not the manufacturer, then a qualifying statement, such as "manufactured for" or "distributed by," must be included. (21CFR101.5)

- Whenever a statement is made as to the number of servings contained in the package, the serving size shall be specified in appropriate units. (21CFR101.8)

- The net quantity of the contents must be given in English units, e.g., quarts or pounds avoirdupois. (21CFR101.105)

- When the labeling or the appearance of the food may create an erroneous impression, the common or usual name shall include a statement of the presence or absence of any characterizing ingredients and/or the need for the user to add such ingredients. (21CFR102.5) Thus, if the label for a frozen "heat and serve" TV dinner illustrates a sumptuous meal of roast turkey, peas, and mashed potatoes garnished with cranberry sauce, but the contents do not include cranberries, then a statement is necessary to the effect that the user must supply his own cranberry sauce.

Conversion to Metric Units

The wording of the FPLA has effectively thwarted the conversion to metric units of all food labeling under the jurisdiction of FDA. Of all the countries in the world, the United States is the last major holdout against metrication, the use of meter, gram, and liter for measurements of length, weight, and volume. Metric units were first adopted around 1792 by the French Republic, which also assumed a decimal time system based on ten hours in a day, 100 minutes in an hour, and 100 seconds in a minute. While the decimal time standard lasted barely eighteen months before being discarded, over the years metrication has gained supremacy.[6] After nearly two centuries the United States is taking the first tentative steps to convert to metrics. In 1975 Congress passed the Metric Conversion Act which established the U.S. Metric Board to coordinate voluntary programs.[7] The Treasury Department mandated that wines be sold in metric bottles by January 1, 1979, and that liquor follow suit twelve months later. Before further progress can be achieved, not only will the FPLA have to be changed, but the National Conference on Weights & Measures, sponsored by the National Bureau of Standards, will have to modify its model code for adoption by the states.[8] But first the proponents of metrication will have to overcome deep-seated resistance to change, as exemplified by one Midwestern supermarket chain which foresees only increased costs from such a conversion.[9]

Federal Trade Commission Act

Former FDA commissioner Donald Kennedy maintained that food advertising including television commercials is an extension of food labeling.[10] However true this relation may be, FTC has been given explicit jurisdiction over food advertising under Section 12 of the Federal Trade Commission Act. This section states, "It shall be unlawful to disseminate any false advertisement by any means for the purpose of inducing the purchase of food." Under the main provision of the Act, which was passed in 1914, the law mandated, "Unfair methods of competition in commerce are hereby declared unlawful." This clause was amended in 1938 to include, "unfair or deceptive acts or practices in commerce." With this change, not only is competition between suppliers regulated, but express protection has been given to consumers as well.[11]

For years FTC carried out the provisions of the Act through piecemeal litigation and consent orders.[12] In 1975, however, with the passage of the Magnuson-Moss Warranty-FTC Improvement Act, FTC was given new authority to promulgate substantive Trade Regulation Rules (TRR) that set standards of conduct and have the force and effect of laws.[13] No sooner was FTC assigned this new power than it opened hearings across the country on a proposed food advertising TRR. Because of the extensive coverage of the subject, the material was split into three parts: Phase I dealt with claims for natural and organic foods, energy and calories, fat as well as fatty acids and cholesterol, health and related matters; Phase II, claims for nutrition, nutrient comparison, nourishment, and foods intended to be combined with other foods; and Phase III, issues of mandatory affirmative disclosure of nutrition information. In addition, separate hearings were scheduled for a TRR on the advertising and labeling of protein supplements.[14]

The public hearings conducted by FTC on food advertising generated voluminous testimony. Realizing the complexity of the issues and admitting that in some cases the benefits would not outweigh the costs, in April, 1980, the Commission dropped plans to proceed with Phases II and III of the proposed TRR on food advertising.[15] Still, FTC was undeterred from going forward with Phase I and took action on rules concerning energy and fats. Under the ruling about energy, advertisements that represent foods as high in energy must disclose that energy means calories. Also, advertisers claiming that a food is useful for weight reduction must disclose the number of calories in a serving. The rule on fats requires disclosure of the amount of fatty acids, cholesterol, and total fat present when claims are made about any one of these substances. Furthermore, the rule prohibits unsubstantiated claims about the relationship between diet and heart disease.[16]

Final rules about natural and organic food claims have been held in abeyance. FTC referred tentative proposals on this subject back to its staff for clarification. In the meantime the regulatory climate began to shift, with more emphasis being placed on voluntary programs.[17] In response to complaints about FTC's activism, Congress passed the FTC Improvement Act of 1980. The most significant provision of this act is that Congress may now veto any final FTC rule before it is promulgated.[18] At the same time, new thinking is astir in FTC about the purpose and need for TRR's. The premise has been put forward that "anecdotal" evidence about the experience of individual consumers is not sufficient grounds for adopting a general regulation. Instead, a TRR needs to be justified on the basis of the frequency with which a problem occurs.[19]

Proposed Legislation

Interest in passing new food legislation never ceases. One of the more significant proposals was Senate Bill S. 641, known as the Consumer Food Act of 1975, which would have required the registration of all plants involved in food processing. By way of enforcement, a new paragraph to Section 403 of the Federal Food, Drug and Cosmetic Act would have classified as misbranded any food "if it was processed in an establishment . . . required to be registered . . . and not so registered."[20] Currently the states have assigned plant numbers to dairy product facilities in accordance with the Federal Information Processing Standards (FIPS), and these numbers must be placed on all labels to identify the producing plant. Another issue concerns the lack of uniformity of state labeling regulations which, except for meat products, need not conform with federal standards. Legislation introduced by Senator George McGovern in 1979 would have provided for the preemption of state and local labeling regulations by FDA.[21]

INGREDIENT LABELING

One of the first lessons learned by the beginning law student is that for every legal principle put forward, invariably there are exceptions to the principle, and not infrequently there are exceptions to the exceptions. Thus, for the rule that ingredients must be listed by their common or usual name on the label in descending order of predominance by weight, numerous exceptions apply. Foremost among the exceptions are standardized foods promulgated under the FD&C Act for which only optional ingredients have to be listed. On the other hand, labels for all meat and poultry products, whether they are standardized or not, must list the ingredients. FDA believes that many consumers are not aware of the prescribed order in which ingredients are placed on the label, and therefore the agency has considered requiring a statement about the fact that ingredients are listed in their descending order of predominance by weight. To date, however, no action has been taken on this proposal.

Incidental Additives

Incidental additives that are "present in a food at insignificant levels and do not have any technical or functional effect in that food" need not be declared on the label. (21CFR101.100) These substances include the following examples:

1. An additive may be incorporated in an ingredient for a functional purpose, but when that ingredient in turn is used in a finished food product, the additive loses its effect. Thus, an anticaking agent added to flour has a technical effect and must be declared on the label of the flour; however, when the flour is used to make a gravy that anticaking additive no longer is relevant and need not be declared on the label for the gravy.

2. A "processing aid" is a substance that is useful in treating a food during a manufacturing step but has no technical effect in the finished product and is present in insignificant amounts. For example, hydrochloric acid or sodium hydroxide used to adjust the pH of a processing stream but subsequently neutralized would not have to be disclosed.

3. Indirect additives, such as those migrating from packaging materials or introduced from food equipment, do not have to be included in the ingredient statement.

FDA has not published a definition for what it considers to be an "insignificant level" of an incidental additive, but informal communication indicates that FDA would look with askance at any label which failed to declare an ingredient which is present in excess of one percent of the weight of the product. Water need not be disclosed on the label of a food to which it is added in order to reconstitute certain ingredients such as dry whey, nonfat dry milk, and powdered eggs. (In a letter to National Food Processors Association, FDA stated that water must be declared in the ingredient statement on labels of reconstituted juices.)[22] When water, however, is added in greater amounts than required to reconstitute such an ingredient to single strength, it must be included in the ingredient list. (21CFR101.4) A marker or tracer used to identify a particular product can go unreported provided permission has been granted by FDA. (21CFR101.100)

Spices, Flavoring, and Coloring

Spices, flavoring, and coloring may be declared by their collective names rather than by individual substances. Particularly in the case of flavorings, of which there are some 1,700 approved compounds and as many as 125 can be used in a single food product, listing the individual flavor substances would be unwieldy if not confusing to the consumer. (FDA issued an advisory opinion on April 30, 1984, that a "flavor enhancer" such as disodium guanylate is not a flavor and therefore must be declared by its common or usual name.[23] A special provision covers monosodium glutamate which must be declared as a separate ingredient. [21CFR101.22{h}{5}]) FDA, however, has gone on record as having the authority to require the listing of any spice, flavoring, or coloring if it is found to cause an allergic reaction among some consumers. The first coloring to require separate listing, effective July 1, 1982, was FD&C Yellow No. 5.

Flavors are defined by FDA as either "artificial" or "natural" depending on their sources and methods of processing. A third class of flavors, "nature-identical," is recognized in Europe by the International Organization of the Flavor Industry. This third category includes substances prepared synthetically or isolated from aromatic raw materials but identical in every way to compounds present in natural products intended for human consumption. "Nature-identical" flavors are approved by the Codex Alimentarius Commission but in the United States are lumped in with "artificial" flavors. If a flavor representation is made for a non-standardized food by word or vignette, then the name of the food must include a flavor statement indicating the presence of any natural or artificial flavoring. If the characterizing flavor is natural but is supplemented or rounded out with other natural flavors, then the statement should include "With Other Natural Flavors," known in the trade as WONF. As an example, a strawberry shortcake product made with insufficient strawberries to alone characterize the product and containing natural strawberry flavoring should be labeled "strawberry flavored shortcake" or "natural strawberry flavored shortcake."[24]

Unlike flavoring, all coloring additives are considered "artificial" regardless of whether they are certified colors derived from coal tars or uncertified color extracts from natural products. To illustrate, if concentrated beet juice is used to impart a reddish color to pink lemonade, it must be labeled as "artificial" coloring. While the use of food colors is not generally restricted, they cannot be employed in deceptive ways or to cover up unwholesomeness. A producer of an eggnog mix in 1981 was issued a regulatory letter advising that the addition of Yellow No. 5 to its product made it adulterated by giving it the appearance of containing more eggs than it actually did.[25] By special provisions, coloring need not be declared in butter, ice cream, and cheese. The rationale for these exceptions is that coloring is added only to compensate for the seasonal variations in appearance. Otherwise, artificial coloring must be indicated as such in the ingredient list or by the name of the coloring additive used, e.g., caramel color.[26,27] By comparison with flavoring and coloring, a "chemical preservative" must be listed by its common or usual name as well as by its generic function, such as "preservative," "to retard spoilage," or "a mold inhibitor." (21CFR101.22)

Collective Names of Ingredients

Many additional ingredients may be declared by their class or collective names as indicated in 21CFR101.4. For example, whey, concentrated whey, reconstituted whey, and dried whey may be listed simply as whey. A standardized food, e.g., mayonnaise, when used as an ingredient in a finished food may be disclosed by its common name followed by a parenthetical listing of its own ingredients. Additional generic or collective names that are permitted are summarized in Table 4.1. This nomenclature has been most controversial in the case of fats and oils.

TABLE 4.1
COLLECTIVE NAMES PERMITTED IN LABELING

Generic Name	Status	Reference
Cultures[a]	Final	21CFR101.4
Dough conditioners[b]	Final	"
Enrichment ingredients[b]	Proposed	FR April 7, 1978, p.14675
Enzymes[c]	Proposed	FCN* Jan 24, 1983, p. 23
Fats and oils[b]	Final	21CFR101.4
Firming agents[b]	Final	FCN* Feb 28, 1983, p. 25
Icing, cake, filling[d]	Advisory opinion	FCN* Jul 4, 1977, p. 36
Leavening agents[b]	Final	21CFR101.4
Minor ingredients[e]	Tentative final order	FCN* Mar 17, 1986, pp.22-23
Preservatives[f]	Final	21CFR101.22(j)
Spices, flavorings, colorings[g]	Final	21CFR101.22(a)
Stabilizers, emulsifiers[b]	Proposed	FCN* July 5, 1982, p. 22
Standardized foods[h]	Final	21CFR101.4
Sulfiting agents[i]	Proposed	FCN* July 14, 1986, pp. 3-8
Yeast nutrients[b]	Final	21CFR101.4

[a]Bacterial cultures may be declared by the word "cultured" followed by the name of the substrate.

[b]The generic name may be used followed by the individual components of the blend in parentheses but not necessarily in descending order of predominance. If one or more of the components is used intermittently this fact may be indicated by the terms "and/or" or "contains one or more of the following."

[c]Enzymes of animal, plant or microbial origin may be declared on cheese labels as "enzymes."

[d]Baked goods containing icing, cake and filling ingredients may list such ingredients separately under the given headings in descending order of predominance.

[e]Minor ingredients present at 2% or less by weight may be listed together at the end of the ingredient statement provided they are preceded by an appropriate phrase, e.g., "contains 2% or less of each of the following:"

[f]Chemical preservatives must be listed by their common or usual names and by a description of their function.

[g]These ingredients may be declared as "spice," "natural flavor," "artificial flavor" or "artificial color" except for FD&C Yellow No. 5 which must be declared as such.

[h]A standardized food used as an ingredient may be listed by its name immediately followed in parentheses by the common or usual names of its components in descending order of predominance.

[i]Sulfiting agents that remain in a food in a significant amount (over 10 ppm) but no longer have a technical or functional effect may be declared by the term, "sulfiting agents."

*Food Chemical News, Washington, D.C.

When a blend of fats or of oils is used as a component of a food product, at a level that does not equal or exceed the most prominent ingredient, the blend may be declared by its collective term, e.g., "vegetable oil shortening," followed by the individual compounds listed in parentheses not necessarily in descending order of predominance. The list may include an oil or fat used only intermittently provided such wording as "and/or" or "contains one or more of the following" is inserted in the ingredient statement. "Hydrogenated" or "partially hydrogenated" should be used to indicate any or all oils so treated.

The method described for declaring fats and oils has several advantages. A food manufacturer is free to adjust the percentages of different oils used depending on availability and market prices without the need to reprint his labels. The use of collective terms also would appear to be more descriptive and therefore comprehensible to the average consumer. Offsetting these advantages is the possibility that the consumer will be deprived of valuable nutritional information.[28] A cover story in *Time* focused attention on the causes of heart disease with implications for diets containing high levels of cholesterol and saturated fats.[29] A label using the "and/or" disjunctive terminology to declare a blend of oils containing palm oil would not reveal the relative abundance of palm oil, which is cheap and high in saturated fats. Canada has gone a step further in collective labeling by proposing to allow all added sweeteners to be listed generically in the ingredient statement as "sugars (sucres)" followed by brackets containing the common names of the sweeteners that are always present and those that may be present.[30]

NUTRITIONAL LABELING

In conjunction with ingredient labeling, nutritional labeling, by supplying data pertinent to widespread health concerns, provides the best profile of a food product. This labeling information is particularly valuable in characterizing newly introduced engineered foods that have been synthesized to meet specific needs. Bacon analogues or "bacon bits" made from soy and other non-meat ingredients typify these novel foods. As greater sophistication is gained in both ingredient and nutritional labeling, the interest in establishing new standards of identity should diminish. Standing in the way of quicker exploitation of nutritional labeling is the still rather rudimentary understanding of nutrition, especially with regard to certain foods like fresh fruits and vegetables. The costs associated with the development of nutritional information acts as a further impediment to the greater use of this procedure. The food industry shares a responsibility with government to perfect nutritional labeling so as to make it more useful to the consumer.

Dietary Guidelines

Nutrition has moved to stage center in food technology. This recognition reflects the accumulated evidence that the American diet is directly accountable for the needlessly high incidence of heart disease, diabetes, tooth decay, high blood pressure, and certain common cancers. Furthermore there is a growing awareness that such macro-nutrients as fat, carbohydrate, protein, salt, and cholesterol are more relevant to today's health problems than the micro-nutrients, namely, vitamins and trace minerals. This conclusion was reached in a report, *Dietary Goals for the United States*, Second Edition, issued December, 1977, by the Senate Select Committee on Nutrition and Human Needs. Essentially the same theme was later expressed in a joint study released February 4, 1980, by the Department of Agriculture and the Department of Health, Education and Welfare. The findings were summarized in seven "Dietary Guidelines for Americans":

1. Eat a variety of foods.
2. Maintain ideal weight.
3. Avoid too much fat, saturated fat and cholesterol.
4. Eat foods with adequate starch and fiber.
5. Avoid too much sugar.
6. Avoid too much sodium.
7. If you drink alcohol, do so in moderation.[31]

Regulations Covering Nutritional Labeling

General interest in nutritional labeling dates back to late 1969 when the White House Conference on Food, Nutrition and Health brought this topic into focus. These discussions led to the issuance by FDA of final regulations in 1973. As adopted, these regulations provide for the optional inclusion of nutritional information on the labels of consumer products except that under two circumstances such labeling becomes mandatory: (1) if the food processor adds any vitamin, mineral, or protein supplement to a product, or (2) if he makes any nutritional claim on the label or in advertising, other than related to sodium content. On the other hand, a statement such as, "For nutrition information write to _____,", with the blank filled in by the manufacturer's name, does not subject the product to the requirements for nutritional labeling. The regulations promulgated by FDA have largely been copied by USDA for meat and poultry products, thus making these provisions all-inclusive.

A standard format must be followed in all nutritional labeling. Nutrient quantities are declared on the basis of the average or usual serving or portion. The serving size as well as the number of servings per container must be given. Calories and the content, expressed in grams, of protein, carbohydrate, and fat must be stated for a given serving. (Metric units have been traditionally used in nutrition and are not restricted by the FPLA in this usage.) In addition, the fatty acid composition and the quantity of cholesterol and sodium may be declared, but disclosures are required if claims are made about them. Recent proposals would make the inclusion of potassium labeling in the nutritional labeling format optional, and sodium labeling compulsory. Protein and seven specified vitamins and minerals, namely, vitamin A, vitamin C, thiamine (vitamin B_1), riboflavin (vitamin B_2), niacin, calcium, and iron must be given as percentages of the U.S. Recommended Daily Allowances (U.S. RDA). Other essential vitamins and minerals, if added to the food, must be included. If a food is commonly combined with another ingredient before eating, e.g., breakfast cereal to which milk is added, nutritional data can be included for the final combination in a second column on

Serving size (l/3 cup, uncooked)		1 oz (28 g)
Servings per container		18

	per 1 oz. cereal	per 1 oz cereal and 1/2 cup vitamin D fortified whole milk
Calories	110	190
Protein	5 g	9 g
Carbohydrate	18 g	24 g
Fat	2 g	6 g
Cholesterol [*]	0 mg/serving	15 mg/serving
Sodium when prepared without salt, not more than [**]	10 mg/serving	60 mg/serving
Potassium	85 mg/serving	270 mg/serving

[*] Information on cholesterol content is provided for individuals who, on the advice of a physician, are modifying their total dietary intake of cholesterol

[**] Does not include sodium from water used in preparation.

Percentage of U.S. Recommended Daily Allowances (% U.S. RDA)

	per 1 oz. cereal	per 1 oz cereal and 1/2 cup vitamin D fortified whole milk
Protein	6	20
Vitamin A	***	2
Vitamin C	***	***
Thiamine	10	10
Riboflavin	***	10
Niacin	***	***
Calcium	***	10
Iron	4	4
Vitamin D	***	10
Phosphorus	8	20

*** Contains less than 2% of the U.S. RDA for this nutrient.

A one oz. serving contains about 0.3 g (1.2%) crude fiber.

FIGURE 4.1. NUTRITIONAL LABELING FOR A BREAKFAST CEREAL

the nutritional label. An example of nutritional labeling for a popular cereal is shown in Figure 4.1.

Under nutritional labeling regulations, a product to which a vitamin, mineral, or protein has been added shall be deemed to be misbranded unless the total content of any such nutrient is at least equal to the label declaration. The level of a naturally occurring or indigenous vitamin, mineral, or protein in a product not fortified with such nutrient must be equal to at least 80 percent of its stated value. A food with a given caloric, carbohydrate, or fat content shall be considered misbranded if that nutrient exceeds 120 percent of its stated value. Any of the following representations will also render a product misbranded:

1. that a food because of the presence or absence of any dietary properties is effective in the prevention, cure, or treatment of any disease or symptom;

2. that a balanced diet of ordinary foods cannot supply adequate nutrition;

3. that the lack of an optimum nutrient content of a food by reason of the soil in which it was grown may be responsible for a diet deficiency;

4. that the storage, transportation, or processing of a food may be responsible for a diet deficiency;

5. that a food has dietary properties when such properties have not been shown to have significant value or need; or

6. that a natural vitamin in a food is superior to an added or synthetic vitamin.

In a major policy shift, FDA in May, 1986, submitted a proposal that would permit health claims on food labels. This reversal was made in cognizance of the growing concern among consumers about health issues. The new policy, however, is not a carte blanche for food processors. It would require that the following conditions be met:

• "The information should avoid any implication that a particular food be used as part of a treatment or therapy oriented approach to health care."

• "The information should be based on and be consistent with widely accepted, and well-substantiated, peer-reviewed scientific data, ..."

• "The information must emphasize the importance of a total dietary pattern."

• "Information on food labeling must not over-emphasize or distort the role of a food in enhancing good health."

• "The use of health-related information constitutes a nutritional claim that triggers the requirements of FDA's regulations regarding nutrition labeling."[32]

U.S. Recommended Daily Allowances

Accurate nutritional labeling requires that valid numbers be used for serving or portion sizes and for U.S. RDA's. In a strategy document published in the *Federal Register*, December 21, 1979, FDA and USDA proposed serving size regulations for some beverages, cereal, and additional product classes.[33] These initiatives, however, are still in limbo. U.S. RDA's are based on nutritional data generated by the Food and Nutrition Board of the National Academy of Sciences — National Research Council. Concern over nutrition goes back at least to the 1930's when President Franklin D. Roosevelt inveighed, "One-third of the nation is ill-housed, one-third ill-clothed, and one-third ill-fed." The National Academy of Science, responding to the need for more nutrition information, published the first edition of *Recommended Dietary Allowances* in 1941. About the same time FDA adopted Minimum Daily Requirements (MDR) for declaring the nutrient values in foods for special dietary use and for dietary supplements of vitamins and minerals. The MDR terminology, however, caused confusion and misunderstanding. In 1972, FDA reported, "The continued existence of the term 'Minimum Daily Requirements' has led to the production, promotion and sale of a variety of dietary supplements that contain large multiples of the minimum daily requirements for vitamins and minerals, and by far exceed the adequate human total daily dietary requirements."[34]

Thus when FDA adopted nutritional labeling in 1973, it substituted U.S. RDA's for the former MDR terminology. The new U.S. RDA's were generally taken as the highest values recommended in the findings of the National Academy of Sciences. The new system, however, has not met with everyone's approval. In 1978 hearings were held in Congress by the House Science and Technology's Subcommittee on Domestic and International Scientific Planning, Analysis, and Cooperation (DISPAC). Its chairman, Representative James H. Scheuer, expressed frustration at the "total dysfunction of our information — of our system for distributing health information."[35] Considerable testimony criticized the misuse by food processors of U.S. RDA's as marketing and promotional devices. One witness for the Council for Responsible Nutrition decried the use of U.S. RDA's for the comparison of fortified food with traditional foods, and he called for limiting the use of U.S. RDA's to describing indigenous nutrients in foods "since only naturally occurring nutrients serve as an indicator of the broader value of food measured."[36] The criticism of U.S. RDA's was summarized as follows:

- they are based on very limited information or on studies of small unrepresentative samples;
- they are intended for groups of people and are not useful for individuals;
- they are limited to the nutritional needs of healthy people;
- they do not cover all the essential nutrients;
- they overstate the nutrient needs of most individuals;
- they do not provide the upper and lower limits or margins of safety and related risks; and
- they are useless to anyone but nutrition experts because they are difficult to understand.[37]

Nutrient Density

Because of the general criticism of the present format used for nutritional labeling, numerous suggestions have been offered to improve it. FDA, USDA, and FTC have encouraged efforts to develop alternative formats and in 1980 contracted with an outside consultant to stimulate new ideas. One of the more appealing concepts is the "nutrient density" principle whereby a nutrient content is reported in milligrams per 100 calories or as a percentage of its U.S. RDA per 100 calories. A variation of this concept was presented as an Index of Nutritional Quality (INQ), which was defined for a given quantity of food as:

$$INQ = \frac{\text{Percent of nutritional requirement supplied}}{\text{Percent of energy requirement supplied}}$$ [38]

Another approach frequently mentioned is to use graphic representations, such as bar graphs[39] or pie charts[40] to present nutrient data. Finally, descriptive ratings, such as excellent, good, fair, poor, or zero, have been suggested.

Definitions for Macronutrients

With the increased interest in macro-nutrients, qualitative terms, for example, "low," "reduced," and "free," have come into general usage to describe the presence or absence of these components. To avoid confusion about the meanings of these terms, FDA has provided definitions for dietetic foods (low- or reduced-calorie) and products that are sodium-free, very low-sodium, low-sodium, and reduced sodium. It also has approved definitions for cholesterol-free, low-cholesterol, and reduced-cholesterol. (Table 4.2) The agency has declined to establish an upper limit for "caffein-free," stating that such a designation cannot be used unless there is no detectable caffein present.[41] Use of such catchy expressions as "light" or "lite" to denote low- or reduced-calorie would come under the regulations.[42] FDA has indicated that mandatory sugar labeling will be held in abeyance.[43] One other macro-nutrient deserves mention, namely, dietary fiber, which has been the subject of much speculation in health articles. "Fiber-enriched" is a term being increasingly used by food manufacturers to convey the

TABLE 4.2
DEFINITIONS FOR QUANTITIES OF MACRONUTRIENTS

Macronutrient	Accepted Term	Definition
Calories[a]	Low calorie	A serving supplies no more than 40 calories, and not more than 0.4 calories per gram
	Reduced calorie	A calorie reduction of at least one-third
Sodium[b]	Sodium free	Less than 5 mg per serving
	Very low sodium	35 mg or less per serving
	Low sodium	140 mg or less per serving
	Reduced sodium	75% or greater reduction
Cholesterol[c]	Cholesterol free	Less than 2 mg per serving
	Low cholesterol	Less than 20 mg per serving
	Reduced cholesterol	75% or greater reduction
Fat[d]	Lean	No more than 10% fat in meat, poultry products
	Extra lean	Not more than 5% fat
	Leaner	25% reduction in fat or greater

[a]21CFR105.66.
[b]*Food Chemical News*, June 18, 1984, p. 23.
[c]*Food Chemical News*, December 1, 1986, pp. 3-8.
[d]*Food Chemical News*, April 7, 1986, pp. 37-39.

impression of good nutrition.[44] Finally, both FDA and USDA have found misleading any negative statement which purports that a food recognized as not normally a source of a given nutrient does not contain the nutrient.[45,46] Under this interpretation, a vegetable oil labeled cholesterol-free, candies represented as salt-free, or ginger ale as caffein-free would be considered misbranded.

FOOD FORTIFICATION

Food fortification refers to the intentional addition of protein, vitamins, and minerals to food in order to improve its nutritional value. This practice has had a long string of accomplishments. As far back as 1833 a French chemist advocated the addition of iodine to table salt to prevent goiter although this idea was not widely accepted until the turn of the century. In 1918 Denmark, recognizing a deficiency of vitamin A in people's diets, provided for the fortification of margarine with this nutrient. The United States around 1931 introduced vitamin D fortification of milk, and in 1941 FDA established a standard of identity for enriched flour that is supplemented with thiamine, niacin, riboflavin, and iron. A recent development in the general area of fortification has been the fluoridation of drinking water supplies to reduce the incidence of dental caries.[47]

Misuses of Fortification

The growing application of food fortification has not been an unmitigated success nor without controversy. Concern has grown relative to the possibility of "over-fortification" whereby some nutrients, typically vitamins A and D, might be consumed, especially by children, in excessive amounts. Another worry is the potential for indiscriminate fortification that could encourage the consumption of "fun" foods which have little redeeming merit from a nutritional point of view.[48] For this reason snack foods like potato chips are considered inappropriate for fortification. FDA has expressed the fear that, unless a rational approach is adopted by the food industry, random fortification could result in "deceptive or misleading claims for certain foods" and "create nutrient imbalances in the food supply." (21CFR104.20)

The public debate over food fortification made national headlines during Senate hearings in 1970 when Robert B. Choate, Jr., a self-styled consumer activist, introduced a chart comparing the nutritional values of sixty popular brands of breakfast cereals. On the basis of this evidence, he charged that two-thirds of these products "fatten but do little to prevent malnutrition," and he described them as supplying "empty calories."[49] Within a year the major cereal companies had increased the fortification of their products, but criticism persisted that cheap chemical fortifiers were being used as substitutes for more expensive natural ingredients and that some pre-sweetened cereals contained more sugar than cereal.

Consumers Union stepped into the fray by undertaking an in-depth study on cereal nutrition. Released in February, 1975, the partial findings were:

> Most of the cereals tested were at least 70 percent carbohydrates, much of it present in the form of sugar, but some as starch. . . . Perhaps to keep sugar from appearing as the main ingredient, some makers use several other sweeteners with it. . . . A cereal can be stuffed with just about every nutrient for which daily requirements have been established and still not be able to support the lives of test animals. And consumers can read cereal labels until they're blue in the face and still not be able to tell how nutritious a given product is. What, then, is the point of using cereals that have been turned for all practical purposes into bulky vitamin pills?[50]

FDA Policy Guidelines for Fortification

Under the present food laws, FDA is powerless to promulgate binding regulations concerning food fortification.[51] Nevertheless, recognizing the need for some guidelines, the agency issued on January 25, 1980, a sweeping policy statement, which, even if it does not have the force of law, certainly should have a major impact by virtue of its purposeful goals and common sense approach. A fundamental premise of this policy is that not all foods are suitable for fortification. Fresh produce, meat, poultry, fish products, sugars, and snack foods such as candies and carbonated beverages are singled out as not being appropriate vehicles for added nutrients. Five situations are presented where fortification may be justified provided certain conditions are met.

• As a public health measure, nutrients may be added to a food to correct a disease caused by a nutritional deficiency among a population group. Action on this principle should be taken only with the advice and consent of FDA.

• Restoration of nutrients lost during storage, handling, and processing is allowed provided good manufacturing practice has been exercised and proper storage and handling procedures have been followed. Full restoration of all lost nutrients is required under this guideline.

• Balanced fortification of fabricated or engineered foods is allowed such that all essential nutrients present (except specified optional ones) are in proportion to the caloric content of the food. Correct nutrient levels of addition are determined by the U.S. RDA's and by the standard daily intake of calories, which has been reduced from 2800 to 2000 kilocalories (usually shortened to calories in nutrition reporting) in cognizance of the sedentary life-style of most Americans.

• A food that replaces a traditional food in the diet may be fortified to avoid nutritional inferiority and consequent labeling as "imitation."

• In compliance with a standard of identity, common or usual name regulation, or a nutritional quality guideline, appropriate nutrients may be added to a food.

Further conditions, in addition to the above guidelines, were laid down for fortification. The added nutrients must be stable in the food under conditions of use, and their bioavailability must not be reduced by the food to which they are added. The level of addition must not contribute to excessive intakes of the added nutrients. Properly fortified foods may be labeled with any claims or statements that are accurate and forthright. Such terms as "enriched," "fortified," and "added" may be used interchangeably to connote the addition of vitamins, minerals, or protein to a food. As an example of an acceptable statement, "vitamins and minerals added are in proportion to caloric content" may be used in accordance with the principle of balanced fortification. (21CFR104.20)

USDA Programs

USDA shares FDA's concerns about fortification and supports the policy set forth by FDA. In assessing the merits of a proposed food fortification, USDA relies on four principles, all of which must be met.

1. There is a demonstrated need for the nutrient in the specified population.
2. The use of the product is not likely to create a dietary imbalance.
3. The vehicle to which nutrients are added is an appropriate one.
4. The use of the product will not be confusing or misleading to consumers.[52]

Notwithstanding these constraints, USDA has sanctioned selective fortification. In its food assistance programs, fortification is sometimes permitted to meet specified nutrient levels. The Supplemental Food Program for Women, Infants, and Children (WIC) as well as the child nutrition and needy family programs permit certain fortified foods. Schools, for example, have made excellent use of these products in their lunch programs.

New Trends in Fortification

What is the future for food fortification? In 1976 ITT Continental Baking Co. reported on the development of a breakfast bar called Astrofood that in conjunction with eight ounces of milk would supply a complete meal. Made by reformulating a creme-filled cake and fortified with nutrients, the bar, when eaten with a glass of milk, offered the equivalent nutrition of four ounces of orange juice, two slices of bacon, one egg, one pat of butter, and one slice of bread. It was tested with some fanfare in school feeding programs. Parents, teachers, and local taxpayers railed at the idea of feeding cake to youngsters although the bar was well received by the students.[53] The product could not make headway against such antagonism and gradually was withdrawn from the market. These results suggest that the best intentioned fortification scheme can be sidetracked by the strong sentiments of traditionalists.

IMITATION AND SUBSTITUTE FOODS

Section 403(c) of the Federal Food, Drug, and Cosmetic Act states that a food is misbranded if it is an imitation of another food unless its label indicates that fact. The law, however, does not provide a definition for "imitation." At first FDA considered the meaning of imitation to include any substitute food, that is, any food that resembles and is intended to replace another food in the diet.[54] Then in an abrupt reversal of policy, FDA in 1973 revised its definition for imitation. The new interpretation, which has been upheld in the courts, states that a substitute food is an imitation food only if it is nutritionally inferior to the food it replaces. Nutritional inferiority is determined by making a comparison of the levels of the essential nutrients, namely, protein, vitamins, and minerals for which U.S. RDA's have been established. Another limitation on substitute foods is spelled out by the Filled Milk Act which prohibits the addition of any fat or oil other than butterfat to milk, cream, or skimmed milk to produce a product with a semblance to these dairy products.

Development of Fabricated Foods

The change in FDA policy on imitation labeling was designed to encourage the development of new fabricated foods. The record has proven that this approach has been eminently successful in fostering the introduction of a host of dietetic, convenience, and budget foods.[55] These substitute foods, because they are nutritionally equivalent to traditional foods, avoid the onus of being labeled imitation. They are required, however, to conform with the regulations for Common or Usual Name for Nonstandardized Foods. A suitable name, which may be a coined term, must accurately identify or describe the new food product. The label may also bear a fanciful name which is not false or misleading. If the food is a dietetic product with a reduction in caloric or fat content, it must comply with the provisions for nutritional labeling.

Restructured foods, which are a special class of substitute foods, may provide cost savings to the producer and the consumer. Examples of such products are onion rings made from diced onions, potato chips made from dried potatoes, and breaded shrimp sticks made from minced shrimp. In each case the characterizing ingredient is cheaper and easier for the manufacturer to process than the intact food item which it replaces. Restructured foods should be labeled with a qualifying statement to indicate that they are made from comminuted ingredients. As in the case of onion rings made from diced onion, this qualifying phrase must immediately follow the product name and be at least half the size of the type used for the name. (21CFR102)

State Regulatory Agencies

Disagreement exists among state regulatory bodies concerning the acceptability of substitute foods. A 1981 survey indicated that 39 percent of the states require "imitation" to be printed on the labels of all cheese substitutes regardless of nutritional equivalency.[56] In the belief that federal regulations are too lax, New York, Wisconsin, and Georgia have led the fight to enact their own imitation labeling statutes.[57] Answers to these conflicting opinions nevertheless are elusive. Take, for example, frozen pudding on a stick — is it a pudding, a frozen dessert, or an imitation ice cream?[58]

Another objection has been raised concerning FDA's regulations for imitation and substitute foods. The regulations exempt any product for which a standard of identity, a common or usual name regulation, or a nutrition quality guideline has been promulgated. Such an automatic loophole allows important new foods that are nutritionally inferior to be marketed on par with traditional foods. To illustrate, a nutritional quality guideline has been proposed for non-carbonated breakfast beverages as obvious substitutes for citrus juices. While it requires nutritional equivalency for vitamin C, it omits any potassium specification.[59] Unfortunately not all oversights may be as easy to spot as this one.

DIETARY SUPPLEMENTS

The outstanding advance of the twentieth century in nutrition was the recognition of the importance of "accessory food factors" or vitamins in the human diet. Christiaan Eijkman's pioneering experiments in 1893-1897 laid the foundation for modern research in nutrition. While living in the Dutch East Indies he observed that beriberi, a disease then prevalent in the Orient, could be cured by ingesting an extract of discarded rice polishings, which subsequently were found to contain thiamine or vitamin B_1. Working with the same nutrient, Casimir Funk in 1912 suggested the name "vitamine," a contraction of vita (for life) and amine (a chemical structure thought to be common to all vitamins), to describe these substances. In general, any substance to be classified as a vitamin must meet two criteria: first, a deficiency of the compound in a diet will produce a characteristic symptom, and second, restoration of the compound will reverse the deficiency symptom.[60] One by one, new vitamins were discovered, their structures elucidated, and quantities of them synthesized for commercial applications. As their production soared and costs plummeted, vitamins became available on a large scale not only for food fortification but as dietary supplements as well.

Commercialization of Dietary Supplements

Manufacturers have been accused of being overzealous in promoting the consumption of vitamins. Closer to the truth is the fact that the public has eagerly sought the real and imaginary benefits of vitamins. These nutrients have assumed a mystique approaching the elixir of life; they are presumed by faddists to be capable of preventing the common cold, treating cancer, providing sexual prowess, and prolonging life. If a small intake of vitamins is beneficial, consumers reason that more must be better. Left unbridled in their beliefs, Americans are swallowing literally billions of vitamin pills each year.

Concerned over the rising trend in vitamin consumption, FDA began investigating the effects of these apparent excesses. One of its first acts was to slap controls on the sale of vitamins A and D, which in large doses were henceforth restricted to prescription drugs. Then in July, 1973, FDA promulgated regulations to limit the daily dosages of vitamin supplements to 150 percent of the established U.S. RDA's. Any preparation in excess of this amount would be regulated as a drug. These rules, which were slated to become effective on January 1, 1975, immediately raised a furor and led to a bitter court case. Rather than continue the fight, FDA succumbed to the opposition, and in March, 1975, the agency retracted its regulations for high potency vitamin and mineral supplements.

Proxmire Amendment to the FD&C Act

Congress, not completely satisfied with this about-face by FDA, was determined to resolve, once and for all, the issue of dietary supplements. In a rider passed at the end of 1975, this legislative body voted in favor of the so-called Proxmire Amendment to the FD&C Act. In earlier debate on the bill, Senator William Proxmire had argued that:

> The health food industry - not the big drug manufacturers, but the health food industry and the small health food stores — the Mom and Pop health food stores found all over this country, are in grave danger of being put out of business. We are proposing this amendment on this bill to save those businesses. We are also doing it to save the consumer from being ripped off, because if vitamins cannot be bought in amounts greater than 150 percent of the recommended daily allowance (RDA) the consumer will have to pay through the nose for quantities of vitamins he now uses routinely.[61]

The Proxmire Amendment, incorporated into the FD&C Act as Section 411, has as its main provisions:

(A) The Secretary may not establish, under Section 201(n), or 401, or 403, maximum limits on the potency of any synthetic or natural vitamin or mineral within a food to which this section applies;

(B) The Secretary may not classify any natural or synthetic vitamin or mineral (or combination thereof) as a drug solely because it exceeds the level of potency which the Secretary determines is nutritionally rational or useful;

(C) The Secretary may not limit, under Section 201(n), 401, or 403, the combination or number of any synthetic or natural (1) vitamin, (2) mineral, or (3) other ingredient of food, within a food to which this section applies.

Exceptions were provided for products represented for use in the treatment of specific disorders, by children under 12 years old, or by pregnant or lactating women. The effect of the Proxmire Amendment is to regulate all oral vitamin preparations as foods unless therapeutic claims are made on the label in which case the product is classified as a drug.

Enabling Regulations

Once Congress had acted, FDA promptly revised the regulations for dietary supplements. New sections 21CFR105.60 and 21CFR105.85, issued in October, 1976, outlined provisions for label statements and standards of identity, respectively. Of primary concern are the following requirements:

- The principal display panel shall bear a conspicuous statement of the usefulness of the supplement.

- The name of the product shall consist of a term which is descriptive of the vitamin and/or mineral composition, together with a phrase designating the group for which the supplement is intended. In addition, a proprietary name may be used so long as it is not false or misleading.

- Except as allowed by law, the potency of each nutrient shall lie within specified lower and upper limits.

- Except as allowed by law, only certain combinations of nutrients are permitted.

- Each nutrient must be listed in a table showing the amount contained in a daily dosage and what percentage of the U.S. RDA that amount is.

- An ingredient statement must give all the active ingredients and excipients.

- An expiration date is required if any of the nutrients is subject to deterioration.[62]

These regulations on dietary supplements, however, were challenged in February, 1978, when a court order remanded these documents to FDA on procedural grounds. The court stipulated that proper notice be given and hearings held prior to rulemaking. As FDA made preparations to re-propose its regulation, it advised that in the meantime manufacturers could use the vacated provisions as guidelines.[63] (Although a proposed regulation does not have the force of law, it still is of judgmental value. FDA has indicated that it generally will not prosecute a business that adopts practices permitted in a proposed rule.[64]) Before the agency could settle this matter, however, a new complication arose.

Over-the-Counter Drug Products

In 1979 a panel, created by FDA to review over-the-counter (OTC) vitamin, mineral, and hematic drug products, issued a highly critical report on the manner in which these products were being promoted. These findings addressed only drug products which comprised about one quarter of the billion dollar vitamin business, the balance consisting of dietary supplements.[65] Even so, the criticism voiced by the panel had a bearing on the sale of all vitamin preparations, and it rekindled old antagonisms. Hoping somehow to reconcile conflicting priorities, FDA in 1981 appointed an agency-wide task force to make recommendations.

In a preliminary report this committee proposed that all oral preparations of vitamins and minerals be treated as foods under the regulations.[66]

Protein Products

Another class of dietary foods, protein products, has been the object of much controversy. A task force, established in December, 1977, by the Department of Health, Education and Welfare to look into the safety of protein products, reported that 60 deaths were associated with the use of these foods as the principal or sole source of calories for rapid weight loss.[67] After almost endless debate in and out of court, FDA on April 6, 1984, published a regulation requiring the following warning statement on protein products deriving 50 percent of their total caloric value from either whole protein, protein hydrolysates or amino acid mixtures:

> WARNING — Very low calorie protein diets (below 400 Calories per day) may cause serious illness or death. DO NOT USE FOR WEIGHT REDUCTION IN SUCH DIETS WITHOUT MEDICAL SUPERVISION. Not for use by infants, children, or pregnant or nursing women.

This regulation was promulgated under Section 201(n) of the FD&C Act which gives as one definition for misbranding the omission of material facts.[68]

NATURAL/ORGANIC PRODUCTS

In our ersatz culture, anything "natural" has an irresistible appeal. Therefore, it is not surprising that food companies whenever possible have been touting their products as "natural." Until recently the meaning of the term "natural" as applied to foods has been fuzzy with the result that its use has sometimes been stretched beyond the point of sound business ethics. Back in 1977 a McCormick & Co. representative observed as "most ridiculous" the term "natural" on a bottle of meat tenderizer, the label of which listed the following ingredients: Sorbitan monostearate, tricalcium phosphate, partially hydrogenated vegetable oil, silicon dioxide, and modified food starch.[69]

Pressed to comment on prevailing industry practices, FDA stated in 1977:

> Explicit policy regarding the use of the term [natural] exists only with regard to products containing artificial colors, artificial flavors, and synthetic ingredients including chemical additives.[70]

After espousing this position, FDA conceded that it was unlikely to bring enforcement action against a producer making "natural" ingredient claims that cannot be substantiated. As an excuse the agency pointed to its "restricted financial resources to pursue such matters."[71]

Involvement by Federal Trade Commission

In the meantime FTC began looking into the usage of the terms "natural," "organic," and "health food" during Phase I of its proposed food advertising trade regulation rule (TRR). In its first attempt to unravel the meanings of these expressions, FTC proposed a two-pronged definition for a "natural" food, which specified that (1) it shall not have undergone more than minimal processing, and (2) it shall not contain any artificial or synthetic ingredient. For a food to be advertised as "organic," FTC proposed that it must have been grown without the application of either chemical fertilizers or synthetic pesticides. When the meaning of "health food" was considered, the FTC staff expressed its opinion that the term was undefinable and recommended that it be banned in advertising except when used in the name of a store.[72]

For several years FTC wrestled with its definitions for natural, organic, and health food. After much deliberation the commission rejected any restriction on the use of the term "health food," and it dropped the definition for "organic" as being unenforceable. The definition for "natural" was confused by such questions as what was meant by "minimal processing" and what effect would such ingredients as refined sugar or bleached flour have on the status of a product. Because of these complications, further considerations of "natural" in Phase I of the TRR was deferred.[73]

Stand Taken by USDA

All along, USDA had taken an adamant stand against the use of the word "natural" on labeling. The first departure from this position came in 1981 when it permitted Louis Rich Foods to call its fresh turkey breast "a natural product." Then in an official policy memo, USDA on November 22, 1982, adopted the proposed FTC definition for a "natural" food as follows:

(1) The product does not contain any artificial flavor or flavoring, coloring ingredient, or chemical preservative (as defined in 21CFR101.22), or any other artificial or synthetic ingredient; and
(2) The product and its ingredients are not more than minimally processed.

USDA reserved the right to determine on a "case by case" basis whether a product qualified as being "minimally processed." It did indicate, however, that the term may include:

(a) those traditional processes used to make food edible or preserve it or make it safe for human consumption, e.g., smoking, roasting, freezing, drying, and fermenting; or (b) those physical processes which do not fundamentally alter the raw product and/or which only separate a whole, intact food into component parts, e.g., grinding meat, separating eggs into albumen and yolk, and pressing fruits to produce juices.[74]

The application of this definition for "minimally processed" very likely will be frustrated by the fact that food processing is continually undergoing subtle changes. The introduction of liquid or condensed smoke to treat meat and poultry products soon posed a question about terminology. This matter was settled on March 22, 1983, by Policy Memo 058 from the Food Safety and Inspection Service of USDA. "Naturally Smoked" was defined as the traditional process whereby the product is exposed to smoke generated by burning hardwoods, corn cobs, etc. "Smoked" was permitted on the labels of products treated with liquid smoke that first has been atomized or vaporized. On the other hand, any product marinated or injected with liquid smoke would have to be identified by the phrase "Natural Smoke Flavor Added."[75]

The second part of USDA's definition for "minimally processed" covers those physical processes "which only separate a whole, intact food into component parts, e.g., grinding meat." This wording immediately brings to mind another process development of some consequence, namely, the separation of meat from bone by mechanical means. Perfected to the point where a minimal amount of bone is entrained in the product, the process still has raised cries of protests from consumers.[76] A final rule issued June 29, 1982, by USDA allows the incorporation of up to 20 percent of such separated meat in popular products like hot dogs and other sausages. The separated meat product has been assigned the term, Mechanically Separated Meat (MSM) as being more descriptive than its former name, Mechanically Deboned Meat. A product containing MSM must be labeled with "Mechanically Separated _____" listed in the ingredient statement but need not include a qualifier in its name.[77]

OPEN DATE LABELING

An open date is a calendar date (as opposed to a coded date which is discussed in Chapter Twelve) and is placed on a food package to indicate to the consumers the freshness of the product. When accompanied by a prefix such as "Packing Date," "Sell By," or "Better If Used By," the open date provides information about when the product was packaged, the last day on which it should be sold, or the day before which it should be consumed. Four common types of open dates have been used in labeling:

Pack Date is the date on which the food was manufactured, processed, or packed. It clearly tells the buyer how old the product is when purchased. For it to be useful, one needs to know something about the expected shelf life of the item. Canned goods and packaged foods, for example, when stored in a cool, dry place generally have extended keeping properties.

Pull or Sell Date is the last recommended day of retail sale, assuming proper storage and handling. It should allow for some holding time in the home before use.

Freshness Date is the date after which the product is not likely to be at peak quality.

Expiration Date is the last date the food should be eaten or used.[78]

Silence of Federal Law

Under federal law open date labeling is voluntary. The only restriction is that if a packer chooses to put an open date on a meat or poultry product, then USDA requires that the date be prefaced by a phrase indicating its significance, e.g., "Use Before." Alternatively, a statement may be used such as "Full freshness 10 days beyond date shown, when stored at 40°F or below." Frozen or refrigerated meat and poultry products must also be labeled "Keep Frozen" or "Keep Refrigerated."

In 1975 the General Accounting Office (GAO) recommended that Congress enact legislation to establish a uniform system for the open date labeling of perishable and semi-perishable foods. GAO pointed out that consumers were confused by the variety of dates (pull date, pack date, etc.) and consequently made only limited use of this information. In 1971 USDA had conducted a study of open dating by a Chicago grocery chain and found that nearly half of the shoppers had no idea concerning the significance of the open dates used. In response to GAO's proposal, FDA and USDA agreed on the need for a uniform system; however, the Commerce Department favored revision of the Model State Open Dating Regulation.[79] To date no action has been taken on GAO's recommendation.

Void Filled by Massachusetts

In this regulatory vacuum, Massachusetts caused a stir by passing in August, 1978, a mandatory open dating statute. The long-debated resolution was slated to become effective July 1, 1979, for most perishable foods; July 1, 1980, for frozen foods; and July 1, 1981, for foods classified as non-perishable including canned products. These regulations immediately met resistance from the National Food Processors Association (NFPA), formerly known as the National Canners Association, which did win some concessions.[80] Subsequently the Grocery Manufacturers of America, the American Frozen Food Institute, and the Sioux Honey Associaion, joined by nine food companies, filed a law suit contesting these regulations. Although these rules were upheld by the Supreme Judicial Court in Massachusetts, a new legal action was taken by General Foods Corporation and Rich-Seapak Corporation, which sought an injunction in federal court on constitutional grounds.[81]

Responding to these objections, Massachusetts passed a revised open date labeling regulation (105CMR520.119) on February 24, 1981. The principal features of this new regulation are:

- Open dating is mandatory for all perishable foods with a shelf life of 60 days or less and for all semi-perishable foods having a shelf life greater than 60 days but less than 90 days.

- An open date must be qualified by either "Sell by Date" or "Best If Used by Date" to indicate its importance.

- Open dating is optional for frozen and long shelf life food products.

- A product shall not be offered for sale after its open date unless it is wholesome, it is segregated from other items, and it is clearly marked as being past its open date.

- Recommended storage conditions should be given if such conditions have a bearing on the open date.

- Exemptions include fresh meat, poultry, fish, fruits, and vegetables; salt and refined sugar; bulk products solely for use in the manufacture of other foods; packaged units of less than 1.5 ounces net weight; and products complying with the open date labeling regulations of another agency.

This regulation by Massachusetts clearly is a milestone in the long saga of open date labeling. Whether subsequently modified, overturned, rescinded, or left intact, these rules should have a bearing on future legislation in the United States.

WARNING STATEMENTS AND INSTRUCTIONS

Although food products may seem innocuous, they are the subject of numerous lawsuits. Cans of pitted cherries may contain an occasional pit that under unfortunate circumstances will be the cause of dental injury.[82] Even more hazardous are certain food ingredients that can expose employees handling these materials to serious occupational dangers. Phosphoric acid, a main ingredient of cola soft drinks, is extremely corrosive and can cause severe burns. If sodium hydroxide, a common processing aid, gets into a worker's eye, critical injury can result. To differentiate ingredients and intermediates from consumer products, it is wise to label the former, "For Manufacturing Use Only."

Product Liability

Product liability is the legal principle which holds the manufacturer of a product responsible for any harm resulting from its use. This tenet, in simplest terms, requires a manufacturer to design and produce the safest possible product, but in the case of an intrinsically unsafe product his duty is to provide adequate warning of the known hazards and instructions on how to use the product correctly.[83] Under doctrines established in common law, a manufacturer can be found guilty in a product liability suit if he is shown to have been negligent; if there was a breach of warranty, express or implied; or if any material fact was misrepresented. In recent years there has been a gradual shift in responsibility which places an even greater burden on the manufacturer to provide a safe product. The principle of strict liability puts the blame squarely on the producer for selling a defective product regardless of the care he may have exercised.[84]

American National Standards Institute

Rules governing warning statements have been formalized for many situations. Under the auspices of the Chemical Manufacturers Association, the American National Standards Institute has issued precautionary label texts for different classes of hazardous chemicals. (The American National Standards Institute, ANSI, was founded in 1918 as a clearing house to coordinate the setting of standards by federal departments and technical/professional societies.) Each statement contains a signal word, such as "danger" or "warning," a statement of the hazard, precautionary measures, and first aid instructions.[85] Because of FDA's earlier concern over hazardous household products, Congress passed the Hazardous Substances Act in 1960. This law, which is now administered by the Consumer Products Safety Commission, provides for the familiar warning, "Keep out of the reach of children" to be placed on all containers of harmful products that could possibly fall into the hands of youngsters.[86]

Department of Transportation

Department of Transportation (DOT) labels or placards are required for shipping any hazardous cargo within the United States. For this purpose hazardous is defined as "any substance which has been determined to be capable of imposing an unreasonable risk to person or property."[87] Under the Hazardous Materials Transportation Act, passed in January, 1975, DOT has established hazard classes, e.g., Class A Explosives, Oxidizer, Corrosive Materials, for which distinctive diamond shaped warning labels have been specified. Practical experience has taught that these warning labels can be printed directly on the product labels to assure they will be affixed to each container and to avoid the application of extra labels. Overseas shipments via air come under the jurisdiction of the International Air Transportation Association (IATA), and shipments by boat are regulated by the Intergovernmental Maritime Consultative Organization (IMCO).[88] A hazardous materials docket issued by DOT in 1980 permits the optional use of United Nations warning labels. Adoption of these labels is part of an effort to harmonize international shipping regulations.[89,90] Bulk shipments require a product identification tag wired to the outlet valve or discharge port of the car or trailer. To prevent its misuse it should be marked, "Remove tag before reloading container."

Occupational Safety and Health Administration

The act creating the Occupational Safety and Health Administration (OSHA) imposes a general obligation on a manufacturer to label all toxic substances stored in his plant, indicating the identity of the material, its potential hazards, and the proper procedures for handling it. High-risk substances for which OSHA has promulgated standards must be marked with specific statements.[91] New rules were issued by OSHA in 1983 to make it incumbent upon manufacturers to provide their workers and their customers with written information about a chemical's hazards. This regulation was hailed by the director of the agency as a "landmark" in OSHA's 12-year history.[92] Industry was largely in favor of this regulation because it promises to reduce the proliferation of worker and community "Right-to-Know" statutes and ordinances throughout the states and their municipalities.[93]

Development Products

The status of products in the development stage with regard to product liability is currently in dispute. There is, for instance, some disagreement in the European Economic Community over whether strict liability covers these products. On this issue Germany stands in the affirmative while France does not have a precedence in support of this position.[94] In the United States the Toxic Substances Control Act (TSCA) specifically excludes "Research and Development" chemicals from the provisions of the law. On the other hand the food laws require that all development products comply fully with the regulations.[95] In light of these circumstances, the prudent approach would be to label trial samples of these products with the statement, "Development Sample — Not for Sale." A disclaimer may be added which states that no warranty, express or implied, is made except as included on the label.

Intermediate Mixes

Intermediate mixes which contain regulated food ingredients require directions for their proper use. To comport with the regulations (21CFR184.1), the label of any such mixture shall include a statement of the concentration of the regulated ingredient or other information to permit the customer to use the product in accordance with good manufacturing practice or any prescribed limitations. Examples of such mixes are those containing gum arabic, guar gum, locust bean gum, gum tragacanth, dill, and garlic.

ALCOHOLIC BEVERAGES

Because of the longstanding interest in the revenues generated by tariffs and excise taxes on alcoholic beverages, the Treasury Department has historically been given responsibility for regulating these products. To help pay the war debts from the American Revolution, Congress passed a tax on whisky in 1791. This duty was unpopular and led to the Whisky Rebellion in western Pennsylvania. The liquor tax was repealed in 1802, and except for a brief period after the War of 1812 it was not reimposed until the Civil War. At this time the Internal Revenue Service was established as a branch of the Treasury Department. This new agency was given authority over alcoholic beverages until 1972 when the Bureau of Alcohol, Tobacco and Firearms (BATF) was created to supervise these and other activities. By 1977 the Treasury was receiving more than eight billion dollars annually from federal excise taxes on alcohol and tobacco products.[96]

Shared Responsibility by BATF and FDA

Since its inception BATF has had a running feud with FDA over which government agency has primary responsibility for regulating alcoholic beverages. FDA has maintained that under the FD&C Act it has authority for all "articles used for food or drink for man or other animals." Disagreeing, BATF claims that the Federal Alcohol Administration Act (FAAA) gives it jurisdiction over malt beverages, wine, and distilled spirits. Differences of opinion were sharpened between the two agencies when consumers began to push for the ingredient labeling of these products.[97]

A Memorandum of Understanding was signed prior to 1975 by BATF and FDA to work out an arrangement for the labeling of alcoholic beverages. Bowing to pressures BATF issued a formal proposal to require ingredient declarations, but it quickly changed its mind and rescinded the proposal. Then FDA, acting on its own initiative, promulgated labeling regulations; however it was rebuffed in a court case brought by Brown-Foreman Distillers Corporation. In an attempt at compromise, BATF on February 2, 1979, put forward a revised regulation that would have permitted the listing of ingredients in other than descending order of predominance.[98] BATF later weakened this proposal by giving manufacturers the option of supplying ingredient information separately upon written request from an interested party. This was a feature which FDA as well as USDA had steadfastly resisted when suggested on numerous occasions for foods.[99] Before even this relaxed proposal was slated to take effect, BATF killed the whole idea of ingredient labeling in a final rule released October 6, 1983. The only exception is the required disclosure of FD&C Yellow No. 5.[100]

Although BATF dropped all ingredient labeling plans, it did order the conversion of container sizes to metric units. Under the law, BATF has also established numerous standards for liquors and wines. It has reviewed the nomenclature for labeling wines and issued opinions on varietal designations (e.g., Chardonnay), appellations, viticultural areas, and the term "estate bottled," but it left "vineyard" undefined.[101] BATF has proposed to define "light" or "lite" for not only wines but also beer and distilled spirits as meaning a reduced caloric content, perhaps with a standard fractional reduction such as one-third.[102] While most wines are being labeled showing the alcohol content, only those containing more than

14 percent by volume are prescribed by statute to give the percentage. Beer is not permitted to be labeled with the alcohol content unless specifically required by state law. (In 1984 BATF approved the use of "Low Alcohol" [LA] on beer labels to indicate an alcohol content of less than 2.5 percent.[103] The popularity of LA malt beverages is part of the trend to "neutered" products, e.g., low-tar cigarettes and decaffeinated coffee.) In a departure from tradition, BATF now requires (allowing for a transition period) that the alcoholic content of distilled spirits be expressed as a percentage by volume rather than degrees proof (100 proof being equal to 50 percent).[104]

Controversy over Warning Statements

Warning labels for alcoholic beverages have been an emotional question. In 1973 a North Carolina man died of acute ethanol poisoning after drinking half a fifth of whisky while in a state of extreme grief for his terminally ill wife. The man's family sought $700,000 in damages plus a requirement that distillers place on bottles a warning that excessive and fast drinking of their products can be fatal. The suit was finally settled in 1977 when the U.S. Court of Appeals for the Fourth Circuit absolved the liquor company of any guilt in the death and ruled that distillers were not required to put health warnings on their products.[105] In 1979 BATF looked into a proposal for a warning statement directed to pregnant women because of reports of fetal alcohol syndrome. Later that same year BATF and FDA were ordered by Congress to investigate the need for warning labels at the time amendments were passed to the 1970 Comprehensive Alcohol Abuse and Alcoholism Prevention, Treatment, and Rehabilitation Act.[106] To date nothing has come of these studies in contrast to the warning labels required on cigarette packages and advertisements.

ANIMAL FEEDS

Animal feeds are an important adjunct of the food industry for three primary reasons. First, many of the ingredients used in feeds are no different from those consumed in human food and are produced in the same plants from identical raw materials. Second, the food industry generates huge quantities of wastes, much of which has been upgraded by concentration, drying, blending, and other means to produce valuable feed components. Third, nonconforming food products on occasion may be diverted to feed applications provided the safety of the animal is assured and deleterious products will not thereby be introduced into the food supply. Although operating in the shadow of the food industry, feed manufacturers have achieved a remarkable record in helping to increase farm productivity. By utilizing the results from advanced nutritional studies, animal health research, and sophisticated economic analyses incorporating linear programing, feed companies have been instrumental in obtaining food utilization efficiencies which were unthinkable a few years ago. Complete animal rations may contain not only the necessary macro-nutrients, vitamins, and trace minerals but also hormones, antibiotics, anthelmintics and other prophylactic agents. Such medicated feeds are designed to promote growth and to prevent disease, particularly in confined areas such as feed lots.

Association of American Feed Control Officials

FDA and the states have dual responsibility over the regulation of feeds. Under the FD&C Act, FDA has promulgated standards for the feed industry, but most of the enforcement has been left to the states. The Association of American Feed Control Officials (AAFCO) has devised a uniform state feed bill and model regulations for adoption by state legislatures and departments of agriculture. Feed operators must file registration papers in the states where they do business and pay the scheduled fees, which are usually based on tonnage figures.[107]

The label format for feeds requires showing the net weight, product name, brand name if any, the name and address of the manufacturer or distributor, and directions for use as needed. A guaranteed analysis of the feed is mandatory. When applicable, percentages should be given for the following nutrient levels: minimum crude protein, maximum equivalent protein from non-protein nitrogen compounds, minimum crude fat, maximum crude fiber, upper and lower limits for minerals, and total sugars as invert. Vitamins should be reported generally in milligrams per pound of feed. Besides the guaranteed analysis, the label must bear a list of feed ingredients. For livestock and poultry feeds, collective terms may be used instead of the common or usual names (21CFR501.110). Thus, corn, barley, grain sorghum, oats, rice, and wheat may be lumped together as "cereal grains." The Pet Food Institute has petitioned FDA for permission to use collective terms for pet foods but was turned down on the basis that these products are consumer items and come under the provisions of the Fair Packaging and Labeling Act. This stand, however, is being reconsidered in a proposal announced January 18, 1983.[108] Accepted ingredient names are listed by AAFCO in its *Official Publication*, although this reference does not pretend to be complete. Additional entries are given in *U.S. — Canadian Tables of Feed Composition*, published by the National Academy of Sciences — National Research Council.

Use of Animal Wastes

In 1980 FDA revoked its ban initiated in 1967 on the use of animal waste in feeds. In making the announcement, FDA said that it planned to rely on the states to monitor this activity.[109] AAFCO in the meantime has established guidelines for the safe use of animal excreta. The recycled material must be free of dangerous levels of pathogens, parasites, pesticides, drugs, and other toxic substances. A warning shall be included on the label if more than one-third of the guaranteed crude protein is supplied by non-protein nitrogen sources, or if the copper content exceeds 25 ppm.

EXPORTED FOODS

The FD&C Act specifies that a product destined for export must be labeled on the outside of the shipping container that it is intended for export, and it must not be sold or offered for sale in domestic commerce. To comply with these requirements, such products might be marked, "For Export Only." In addition they most likely should be labeled, "Made in U.S.A." to meet the requirements of country-of-origin statutes on the books in foreign countries. As far back as 1887, England, besieged by a flood of imported industrial goods, passed a law requiring that imported products be labeled showing where they were made.[110] The same condition holds true for products being shipped to the United States. The Tariff Act of 1930 directs that the English name of the country of origin be indelibly marked on every imported article. FDA and USDA have expressly supported this policy for all consumer packages.[111] Importers of food into the United States must follow special procedures to expedite the entry of these goods.[112]

Foreign Food Laws

Although food laws vary from country to country, the general tenor of these regulations is similar. England's Food and Drugs Act of 1955 states that a person who supplies a label which "(a) falsely describes the food or drug, or (b) is calculated to mislead as to its nature, substance or quality, shall be guilty of an offence. . . ."[113] Because of the common goals and objectives among foreign food laws, attempts to introduce international standards are making steady progress. The European Economic Community (EEC) has formulated a regulation for the labeling, promotion, and advertising of all consumer food products. The following particulars, which are amplified in other sections of the regulation, must be indicated on all labels:

1. The name under which the product is sold.
2. The list of ingredients.
3. In the case of prepackaged foodstuffs, the net quantity.
4. The date of minimum durability.
5. Any special storage conditions or conditions of use.
6. The name or business name and address of the manufacturer or packager, or of a seller established within the Community.
7. Particulars of the place of origin or provenance in the cases where failure to give such particulars might mislead the consumer to a material degree as to the true origin or provenance of the foodstuff.
8. Instructions for use when it would be impossible to make appropriate use of the foodstuff in the absence of such instructions[114]

Members of the EEC have moved forward to adopt the regulation, specifying that all labeling be in their national languages.[115]

Impediments to Trade

Notwithstanding progress on coordinating food regulations, differences persist in various countries of the world. The Scandinavian states of Norway and Sweden have restricted the use of food colors in their products. Norway does not allow synthetic colors, and Sweden forbids any added coloring in infant foods, milk products or milkshakes, butter, vanilla ices, fruit, fish and meat products.[116] Canada passes on the functionality as well as the safety of food additives before approving their use.[117] Canadian consumer products must abide by the provisions of the Consumer Packaging and Labelling Act that has many parallel features to its American counterpart, the Fair Packaging and Labeling Act.

Even though Japan is dependent on other nations for large quantities of foodstuffs, it has traveled an independent course in food regulation. The "Lemon War" broke out between Japan and the United States in 1975 when Japan seized large shipments of imported lemons from the U.S. This fruit contained residues of a fungicide, orthophenylphenol, which was approved in all major countries of the world except in Japan. After much haggling this issue was resolved to the benefit of both countries.[118] With more prodding Japan is easing other barriers to trade. It has translated into English its food regulations including labeling standards for distribution to shippers.[119] Once the idiosyncrasies have been grasped, Japan has been found to be an excellent market that is well worth the effort to cultivate.

REFERENCES

1. "Advertising Guidelines for Dealing with Technical Issues Regarding Food Safety and Nutrition," *Food Technology*, November, 1978, p. 32.
2. "Food Labeling, Public Hearings," *Federal Register*, June 9, 1978, pp. 25296-25308.
3. Stephanie Crocco, "FDA's New Era . . . Expressway to Chaos or Sound Policy?" *Food Engineering*, July, 1976, pp. 60-62.
4. "How USDA Evaluates Packaging Materials," *Food Engineering*, March, 1978, pp. 111, 112.
5. "Fair Packaging and Labeling Act: Law Without Teeth?" *Food Engineering*, February, 1978, pp. 91, 92, 94.
6. Richard A. Carrigan, Jr., "Decimal Time," *American Scientist*, May-June, 1978, pp. 305-313.
7. John H. Jennrich, "The March Toward Metrics," *Nation's Business*, September, 1978, pp. 95-97.
8. "The Long Road to Metric," *Packaging Digest*, June, 1978, p. 14.
9. Robert Levy, *Dun's Review*, January, 1980, pp. 49, 51.
10. *Food Chemical News*, October 31, 1977, p. 14.

11. Federal Trade Commission, *Your FTC — What It Is and What It Does*, Washington, D.C.
12. Elaine S. Reiss, "Advertising and Product Development," *Food Technology*, July, 1977, pp. 75-77.
13. Mel Seligsohn, "Is the FTC on a Holy Crusade against the Food Industry?" *Food Engineering*, February, 1977, pp. 20, 21, 24.
14. *Food Processing*, May, 1976, pp. 23, 26.
15. "FTC Notice Proposes Withdrawal of Phase II, III of Food Advertising Rule," *Food Chemical News*, April 7, 1980, pp. 38-40.
16. "FTC Issues Final Food Ad Rule," *Food Engineering*, July, 1980, p. 42.
17. Mel Seligsohn, "FTC Vs. the Food Industry: Is a New Era Dawning?" *Food Engineering*, November, 1980, pp. 60-62.
18. "New FTC Law Clarifies Mandate," *Food Engineering*, August, 1980, pp. 40, 42.
19. "Muris Hits 'Anecdotal Evidence' in Recommending that FTC Abandon Food Rule," *Food Chemical News*, June 21, 1982, p. 42.
20. Mary T. O'Brien, "Self-Generated Safety Assurance Dictated by S. 641," *Food Product Development*, December, 1975, pp. 68, 71, 72.
21. "McGovern Food Bill Seeks Flexibility for Food Industry," *Food Engineering*, September, 1979, pp. 41, 45.
22. *Food Chemical News*, April 2, 1984, p. 7.
23. *Food Chemical News*, May 7, 1984, pp. 12, 13.
24. Robert C. Lindsay, "Flavor Ingredient Technology," *Food Technology*, January, 1984, pp. 76-81.
25. "Yellow No. 5 Use Denied for Egg Nog Mix," *Food Chemical News*, October 12, 1981, p. 42.
26. Ann Przbyla, "Colors," *Processed Prepared Food*, September, 1980, pp. 109-114.
27. "FDA Issues Advisory Opinion on Listing Color Additive Ingredients," *Food Chemical News*, March 17, 1980, pp. 17-18.
28. "FDA Grants Labeling Exemption for Firming Agents," *Food Chemical News*, February 28, 1983, pp. 25-27.
29. Claudia Wallis, "Hold the Eggs and Butter," *Time*, March 26, 1984, pp. 56-63.
30. "Canada Proposes Changes in Labeling Requirements," *Food Technology*, December, 1980, p. 85.
31. Jane E. Brody, "U.S. Acts to Reshape Diets of Americans," *The New York Times*, February 5, 1980, pp. A1, B16.
32. "FDA to Require Health Claims Be 'Consistent' with 'Substantiated' Data," *Food Chemical News*, May 26, 1986, pp. 3-9.
33. "USDA to Propose Serving Sizes for Various Foods," *Food Chemical News*, December 24, 1979, p. 17.

34. Chris Lecos, "REA's: Key to Nutrition," *FDA Consumer*, November, 1982, pp. 24, 25.
35. "RDA's On Trial," *Food Engineering*, August, 1978, pp. 22, 23, 24, 26.
36. Harold V. Semling, Jr., "How to Make the RDAs Meaningful for Consumers," *Food Processing*, September, 1978, pp. 20, 22, 24.
37. "RDA's On Trial," op. cit.
38. Bonita W. Wyse et al., "Nutritional Quality Index Identifies Consumer Nutrient Needs," *Food Technology*, January, 1976, p. 22.
39. "R & D Trends," *Food Development*, September, 1981, p. 18.
40. "Agencies Labeling Activity Heats Up, Kennedy Prods Industry on Labeling & Ads," *Food Engineering*, June, 1978, p. 28.
41. "FDA Denies 'Caffeine Free' Label for Cola Containing 1 P.P.B. Caffeine," *Food Chemical News*, May 31, 1982, p. 34.
42. Louise Fenner, "That Lite Stuff," *FDA Consumer*, June, 1982.
43. "Mandatory Sugar Labeling is Dead, Hayes Indicates," *Food Chemical News*, June 7, 1982, p. 7.
44. James L. Vetter, "Fiber as a Food Ingredient," *Food Technology*, January, 1984, pp. 64-69.
45. *Food Chemical News*, March 31, 1980, p. 29.
46. "USDA Policy Memo Outlines Negative Ingredient Labeling Guidelines," *Food Chemical News*, February 9, 1981, p. 26.
47. Gilbert A. Leveille, "Food Fortification — Opportunities and Pitfalls," *Food Technology*, January, 1984, pp. 58-63.
48. "Supplementation of Foods Vs. Nutrition Education," *Food Technology*, July, 1974, pp. 55-63.
49. "What's for Breakfast?" *U.S. News & World Report*, August 17, 1970, p. 45.
50. "Which Cereals Are Most Nutritious?" *Consumer Reports*, February, 1975, pp. 76-82.
51. Stephen H. McNamara, "FDA Concerned about Nutrition of Engineered Foods," *Food Engineering*, June, 1977, pp. ef-16, ef-17, ef-20.
52. Mary C. Jarratt, "USDA's Position on Food Fortification," *Food Engineering*, May, 1982, pp. 123-124.
53. Richard D. McCormick, "Product and Process Responses to a Moving Nutritional Target," *Food Product Development*, May, 1976, pp. 60, 62.
54. "FDA, USDA Ask Comment on Options for Imitation Labeling," *Food Chemical News*, December 24, 1979, pp. 14, 15.
55. Karen Moore, "Labeling, Image Problems Plague Substitute Foods, but Astute Marketing Plans Breed Successes," *Food Product Development*, June, 1979, pp. 12, 14, 18.
56. Truman Graf, "Imitations Giving Natural Cheese a Strong Competitive Tussle," *Dairy Record*, July, 1981, pp. 94-95.

57. *Food Chemical News*, December 20, 1982, pp. 41, 42.
58. Jay Sandler, "Pudding: New Star for Dairies?" *Dairy Record*, June, 1981, p. 7.
59. H. P. Sarett, "Inconsistencies in FDA's Application of the Term 'Imitation'," *Food Product Development*, February, 1976.
60. The Editors of Consumer Guide, *The Vitamin Book*, Simon and Schuster, New York, 1979, p. 36.
61. "FDA Vitamin Curbs Rejected in Senate by Decisive Margin," *Chemical Marketing Reporter*, September 30, 1974, pp. 5, 25.
62. *Federal Register*, March 15, 1977, pp. 14329-34.
63. "FDA to Publish Revised Dietary Regulations as Proposals," *Food Chemical News*, April 23, 1979, pp. 6, 7.
64. "FDA Finalizes Rule on Fat Content Disclosure," *Dairy Field*, July, 1980, p. 11.
65. Annabel Hecht, "Vitamins Over the Counter: Take Only When Needed," *FDA Consumer*, April, 1979, pp. 17-19.
66. "Task Force Says Bureau of Foods Should Take Over Vitamin-Mineral Jurisdiction," *Food Chemical News*, November 2, 1981, p. 18.
67. *Federal Register*, April 4, 1980, pp. 22904-14.
68. "Language of Protein Product Label Statements Revised by FDA," *Food Chemical News*, April 9, 1984, pp. 8-15.
69. "Definition Coming for 'Natural Foods'?" *Quality Progress*, September, 1977, p. 7.
70. "FDA Has Not Defined 'Natural' for Articles of Food or Drink," *Food Chemical News*, October 10, 1977, pp. 43, 44.
71. "FDA Says It Is Unable to Take Action Against 'Natural' Claims," *Food Chemical News*, October 20, 1980, p. 34.
72. "FTC Staff Sets Definitions for 'Natural,' 'Organic,' Bans Term 'Health Food'," *Food Chemical News*, December 4, 1978, pp. 9, 10.
73. "FTC Defers 'Natural' Issue; Approves Much of Staff-Proposed Phase I Food Rule," *Food Chemical News*, May 26, 1980, p. 33.
74. "USDA Adopts Proposed FTC Definition of 'Natural' for Label Claims," *Food Chemical News*, December 13, 1982, pp. 20, 21.
75. "FSIS Sets Policy for Smoked Products," *Food Chemical News*, March 28, 1983, pp. 15, 16.
76. *Food Engineering*, November, 1976, pp. 20, 90, 92, 93, 96, 97.
77. "USDA Retains 20% Limit on Mechanically Separated Meat Use in Products," *Food Chemical News*, July 5, 1982, pp. 3, 4.
78. Margaret Morrison, "A Consumer's Guide to Food Labels," *FDA Consumer*, June, 1977, pp. 4-7.
79. "GAO Urges Uniform Open Dating Legislation," *Food Chemical News*, February 10, 1975, pp. 7, 8.
80. "Questions on Label Bills," *Packaging Digest*, September, 1978, pp. 10, 12.

81. "Massachusetts Open Dating Rule Challenged by Two Firms in Court Suit," *Food Chemical News*, February 25, 1980, p. 28.
82. Gordon M. Betz, "Product Lawsuits," *Food Engineering*, July, 1980, pp. 83-87.
83. "Product Liability Model Law Drafted," *Chemical & Engineering News*, January 22, 1979, pp. 18, 19.
84. Charles E. Witherell, "The Products-Liability Threat," *Chemical Engineering*, January 24, 1983, pp. 72-87.
85. American National Standard Institute, *ANSI Bulletin*, Z129.1-1976, January 15, 1976.
86. *FDA Consumer*, May, 1978, p. 26.
87. Ken Snow, "Shipping Hazardous Material," *Chemical Engineering*, November 6, 1978, pp. 102-108.
88. "Know Law's Stiff Demands," *Package Engineering*, January, 1979, pp. 58-60.
89. "Carriers Complain about Hazardous-Cargo Rules," *Chemical Week*, July 16, 1980, pp. 69, 70.
90. Alan S. Brown, "Shipping Hazardous Materials: The Transatlantic Rules Gap," *Chemical Business*, September 22, 1980, pp. 16-23.
91. Edwin F. Vandergrift, "Meeting OSHA Regulations on Toxic Exposure," *Chemical Engineering*, June 2, 1980, pp. 69-73.
92. "Show-and-Tell Time for Hazards," *Chemical Week*, November 2, 1983, pp. 13, 14.
93. "Right to Know: 'Hottest' New Pressure for Industry," *CMA News*, February, 1983, pp. 5-8.
94. "The Widening Shadow of Product Liability," *Chemical Week*, February 3, 1982, pp. 44-48.
95. "FDA Affirms Denial of Bakery Labeling Petitions," *Food Chemical News*, July 23, 1979, p. 62.
96. Frank J. Kreysa and Peter C. Buscemi, "Revenooers in Lab Coats," *Chemtech*, March, 1977, pp. 146-152.
97. "FDA Says Alcoholic Beverages Must Comply with Ingredient Labeling Laws," *Food Engineering*, February, 1976, p. 11.
98. "BATF Ingredient Labeling Proposal Does Not Require Predominance," *Food Chemical News*, February 5, 1979, pp. 21-27.
99. "BATF May Give Alcoholic Beverage Option on Ingredient Labeling," *Food Chemical News*, May 26, 1980, pp. 3, 4.
100. "Ingredient Labeling of Wine, Spirits," *Food Engineering*, December, 1983, p. 13.
101. "Wine Labeling — Final Rule," *Food Processing*, October, 1978, p. 14.
102. " 'Light' Alcoholic Beverages Nutrient Labeling Re-proposed," *Food Chemical News*, August 18, 1986, pp. 25-30.
103. *Food Engineering*, September, 1984, p. 30.

104. "Alcohol by Volume Content Labeling for Distilled Spirits Adopted," *Food Chemical News*, October 13, 1986, p. 38-40.
105. "Alcoholic Beverage Laws," *Food Engineering*, September, 1978, p. 127.
106. "Congress Asks BATF, FDA Decision on Booze Warnings by June 1, 1980," *Food Chemical News*, December 24, 1979, p. 78.
107. *Official Publication*, Association of American Feed Control Officials, 1984.
108. "Seeks Comments on Pet Food Labeling," *Food Technology*, March, 1983, p. 44.
109. "FDA to OK Feed Use of Recycled Animal Waste, with State Regulation," *Food Chemical News*, January 28, 1980, p. 66.
110. *Technology Review*, November/December 1982, p. 50.
111. *Food Chemical News*, December 24, 1979, p. 19.
112. *Importing Foods into the United States*, DHHS Publication No. (FDA) 81-2142, Food and Drug Administration, Washington, D.C., 1981.
113. A. G. Ward, "Safeguarding Our Food," *Nutrition Review*, May, 1977, pp. 116-121.
114. *Official Journal of the European Communities*, No. L33, February 8, 1979, pp. 1-14.
115. "Europe's Labels on Foods Move toward Accord," *Food & Drug Packaging*, March, 1982, pp. 1, 47.
116. "Far-Reaching Scandinavian Color Regulations Outlined," *Food Chemical News*, April 20, 1981, pp. 22, 23.
117. H. C. Grice, "Food Additive Evaluation and Regulation: the Canadian Approach," *Food Product Development*, May, 1977, pp. 28, 30, 32.
118. Henry B. Fayans, "Impact of Food Laws, Regulations, and Standards on World Food Supply," *Food Technology*, June, 1979, pp. 76, 77.
119. "U.S. Pushes Japan on Easing Barriers on Imports of U.S. Food," *Food Engineering*, July, 1982, pp. 16, 21.

PART II

Manufacturing

CHAPTER FIVE
GOOD MANUFACTURING PRACTICE

The FDA regulation known officially as Current Good Manufacturing Practice (or simply referred to as GMP's) establishes manufacturing standards for food to ensure its safety and wholesomeness. The regulation covers all aspects of food processing including employee training, design and construction of facilities, maintenance, sanitation, operations, testing procedures, and record-keeping. The concept of good manufacturing practice has already been introduced in Chapter Three. GRAS substances not otherwise restricted must be used according to good manufacturing practice. The principle again is invoked as a condition for setting tolerances for unavoidable contaminants pursuant to Section 406 of the FD&C Act. The underlying tenet in both situations was aptly expressed by the authors of *Panic in the Pantry*. Their conclusion was: "We're coming to accept the concept that nothing is 'completely safe.' There are only safe ways to use substances."[1]

The key word, often overlooked, in the title of the GMP regulations is "Current." This modifier recognizes the fact that manufacturing standards are not static but are continually undergoing refinement. Furthermore, these standards are relative, not absolute. Thus, performance is measured against the best that industry can do, not on the basis of some utopian goal. Or considering the situation another way, if a manufacturer is doing a commendable job in quality assurance but someone else's results are better, then there is obvious room for improvement. In these circumstances, government regulators must appreciate that unnecessary codification of the GMP's can only serve to frustrate attempts to keep these standards current.[2]

One set of regulations does not suffice for the processing of all foods. Because of their special characteristics, certain foods must be processed under different good manufacturing practices. For example, USDA has promulgated specific sanitary standards for meat, poultry, and egg products. After a serious breach in enforcement, FDA has adopted a new quality control regulation for infant formula. Lastly, milk and shellfish are regulated under separate ordinances which are designed to prevent the spoilage of these foods and ensure their safety.

UMBRELLA GMP'S

The expression Current Good Manufacturing Practice was first put forward in the 1962 Kefauver- Harris Amendment to the FD&C Act. This amendment, which applied only to drugs, stated that a product would be considered adulterated if "the methods used in, or the facilities or controls used for, its manufacture, processing, packing, or holding do not conform to or are not operated or administered in conformity with current good manufacturing practice." This provision, however, did not attempt to define current good manufacturing practice so that the responsibility fell on FDA to interpret its meaning in subsequent regulations.

Promulgation of Umbrella GMP's

In 1964 the National Academy of Sciences-National Research Council advanced the notion that GMP's be applied to the manufacture of food as well as drugs. FDA proposed the first GMP's for food in 1967, and after reviewing numerous comments and making significant changes in the proposal, the agency issued a final rule on April 26, 1969.[3] This regulation, which later came to be known as the "umbrella" GMP's, set forth general principles to be followed in the manufacture of all food products. Initially there was some debate whether this regulation was to be considered as a set of guidelines or whether it was to have the force of law. FDA insisted on the latter interpretation, and in support of its position, it quoted from Section 402 of the FD&C Act:

> A food shall be deemed to be adulterated . . . . if it has been prepared, packed, or held under insanitary conditions whereby it may have become contaminated with filth, or whereby it may have been rendered injurious to health.

The significance of this provision cannot be stressed too much. Insanitary conditions in a plant or warehouse automatically render all food produced or stored in these locations adulterated whether or not the food is actually contaminated with filth or other extraneous matter.

Adoption of Specialized GMP's

Soon after publication of the umbrella regulation, FDA embarked on an ambitious program to investigate the manufacturing requirements of the various segments of the food industry. For each of the product categories, the agency planned to issue specialized GMP's to be appended to the umbrella regulation. In short order GMP's were published for smoked fish(1970), frozen raw breaded shrimp (1970), low-acid canned foods (1973), and cacao products and confectionery (1975). (The GMP's for smoked fish were revoked in 1984[4], for frozen raw breaded shrimp in 1986[5], and for cacao products and confectionery in 1986.[6]) Regulations have also been promulgated for acidified foods and for bottled drinking water.

Under the Safe Drinking Water Act passed in 1974, the Environmental Protection Agency (EPA) was assigned responsibility over "drinking water," but in a subsequent Memorandum of Understanding (MOU) between EPA and FDA, the latter agency assumed responsibility for bottled water and water used as an ingredient in food products.[7]

At one time over twenty separate GMP regulations were contemplated.[8] The futility of this endeavor soon became apparent. Not only did this undertaking require a major investment in time and personnel, but much of the effort became repetitious. Therefore, FDA decided that the preferred approach would be to update and expand the umbrella GMP's to incorporate as much specialized know-how as would have been included in most of the planned regulations.

Revisions in the Regulations

A proposal to revise the umbrella GMP's was published in the *Federal Register* in 1979. It was far more detailed in specific instructions than the regulation on the books. For example, instead of specifying that temperature must be carefully monitored, the new proposal stated that perishable foods had to be refrigerated at 45 °F or below, frozen foods stored at 0 °F or below, and hot foods maintained at 140 °F or above.[9] Public hearings were held in Chicago, San Francisco, and Atlanta to obtain comments on the proposal. After an extended review period, the proposed revision was finally dropped in 1982 at the urging of the Office of Management and Budget (OMB). Prior to this decision, FDA Commissioner Jere E. Goyan in January, 1981, had adopted a new policy which had been recommended by an ad hoc GMP Regulation Task Force. The gist of the policy allowed that "to the extent possible, GMP regulations should be constructed to address 'what' is the requirement and not specify 'how' the requirement is to be achieved." The policy went on to state:

> The FDA will promulgate GMP regulations for an industry or an identifiable segment of an industry when the consequence of a manufacturing failure of that industry's product would be a public health risk. Further, GMP regulations may be promulgated when manufacturing failures or violative conditions other than health risks are commonly encountered and could be prevented or avoided if certain minimum manufacturing or processing practices were prescribed.
>
> Each of the requirements of a GMP regulation shall be designed to prevent or avoid offering a product for sale which might be defective (violative) in a significant respect. Further, the requirements of a GMP regulation shall be focused on prevention of those risks (violations) that cannot be adequately prevented by other approaches such as industry education and publication of industry guidelines.[10]

In the spirit of this policy, FDA reissued the umbrella GMP's in 1986 with only slight changes from the original. Proposals for extensive record keeping and coding of product were dropped as being too costly. (See Chapter Thirteen) On the other hand, a new provision was included which specifies the installation of thermometers to aid in the protection against microorganisms. An economic survey revealed that the provision for thermometers would cost industry less than $1 million per year. This amount was considered to be reasonable and within the constraints imposed by OMB.[11]

INDUSTRY QUALITY ASSURANCE ASSISTANCE PROGRAM

The specifics of the umbrella GMP regulation (21CFR110) should form the basis for periodical plant inspections. A checklist, such as the one illustrated in Table 5.1 can be prepared to highlight those topics which should be covered. Such a list should be modified and expanded to reflect the needs of the operation for which it is intended. These changes should be based on any relevant specialized GMP's, regulations as may be promulgated by the United States Department of Agriculture (USDA), industry guidelines, and, not to be overlooked, the experience of the manufacturer. This exercise is not academic. FDA has prepared its own checklist tied to the GMP's and included it in the *Inspection Operations Manual*. Before an FDA inspector visits a plant, he is primed with this material.[12]

Vast Differences between Products

There are appreciable variations in manufacturing practices between the different segments of the food industry. To cite a few examples, a producer of an insensitive ingredient, such as phosphoric acid, clearly does not have to prescribe hair nets and caps for its workers. Also, much of the equipment may be exposed to the open without jeopardizing sanitation. In another food category, a chocolate manufacturer will not clean the food-contact surfaces of his equipment in the conventional way using wet methods because to do so would only compound his sanitation problems. Instead, he will flush out his system with a material that is compatible with the product being produced. The principal concerns of the processor of smoked fish are quite different. In this case, particular attention must be paid to maintaining the proper concentration of salt and controlling the temperature at specified levels in order to minimize the hazard of botulism.

No manufacturing specifications are more critical than those dictated in the GMP's for low-acid canned foods, defined as having a pH above 4.6 and water activity greater than 0.85. The reason is that lack of control at the retorting step exposes the product to the danger of botulism. Elaborate procedures, therefore, have been worked out for determining, controlling, and recording the critical variables. FDA's surveillance over canning operations is the most stringent in the food industry. If significant deviations are noted, the processor is required to obtain an Emergency Permit to avoid having his plant shut down. Such a permit will allow an operator to stay in production on a probationary status until FDA is satisfied that corrective actions have been taken. The Emergency Permit Control regulation (21CFR108) was first applied in 1974 under Section 404 of the FD&C Act.[13]

Alternatives to Specialized GMP's

Most of the specialized GMP's are no longer in force, and none are proposed or in preparation. To fill this regulatory gap, FDA is turning to the food industry for assistance. Beginning in 1968, the agency organized on a pilot basis the Cooperative Quality Assurance Program (CQAP) in which food companies were invited to participate. Under this program a continuous reporting system was established for selected plants which voluntarily kept the government informed of detailed operating results, including all deviations in periodic Exception Reports.[14] This self-certification program was extremely effective in spotlighting quality problems on a plant by plant basis, but after ten years of trial, FDA concluded that CQAP did not permit the most efficient use of the agency's limited resources.

The most recent attempt to enlist industry support, and one which has promise for the future, is the Industry Quality Assurance Assistance Program (IQAAP). This program fosters a working relationship between FDA and an industry association with the purpose of developing standards for use by members of the association. Shortly after the inception of the program in 1980, the Retail Confectioners International and the National Peanut Council both expressed interest in joining.[15] Perhaps the most active participant to date has been the American Institute of Baking (AIB), which is seeking to develop a comprehensive quality assurance plan for the baking industry that addresses the requirements of both the smaller retail baker and the larger bakeries that sell to supermarkets.[16] AIB brings to this project a wealth of knowledge accumulated through its affiliate, the Baking Industry Sanitation Standards Committee (BISSC). This committee was founded in 1949 by six national organizations serving the baking industry by developing standards regarding the design and construction of baking equipment.[17] Although largely exploratory at this stage, IQAAP needs to resolve questions of procedure as it expands. The program might consider adapting to its own purpose the principles contained in FDA's administrative practices and procedures:

> When a trade association files an objection or request for hearing in a proceeding that permits an opportunity for a formal evidentiary public hearing, all subsequent action by the association with respect to such matters binds each member except to the extent that that member independently files its own objection or request for hearing or is otherwise specifically excluded from representation by the trade association in that matter, in which case its rights shall be entirely separate and distinct.[18]

TABLE 5.1
PLANT INSPECTION CHECKLIST FOR GMP'S*

PERSONNEL
1. Are workers with communicable diseases quarantined?
2. Are outer garments clean and tidy?
3. Do employees wash their hands before returning to work?
4. Are all insecure jewelry and personal effects removed?
5. Are correct gloves, hair nets, and caps worn?
6. Are coffee, lunch, and smoking breaks taken in removed areas?
7. Are all personal care products and belongings kept out of food handling areas?
8. Do workers pay attention to personal hygiene?
9. Are routine training programs held?
10. Are supervisors aware of their responsibilities?

PLANT AND GROUNDS
11. Is the immediate vicinity of the plant free from refuse, brush, and junk?
12. Are roads and parking lots paved?
13. Are grounds properly drained?
14. If neighbors present any hazards, have corrective steps been taken?
15. Does the equipment layout allow easy access and room for maintenance and cleaning?
16. Are floors, walls, and ceilings in sanitary condition?
17. Are fixtures, ducts, and pipes suspended away from working areas?
18. Are aisles and working spaces unobstructed?
19. Are food handling areas effectively separated or partitioned from the rest of the plant?
20. Is lighting adequate and enclosed in safety fixtures?
21. Is there sufficient ventilation?
22. Are plant openings effectively screened or covered?

SANITARY FACILITIES AND CONTROLS
23. Is the water supply safe?
24. Is there proper sewage disposal?
25. Is plumbing correctly sized and installed to prevent contamination of food products?
26. Are toilet and hand washing facilities clean?
27. Is there proper rubbish disposal?

SANITARY OPERATIONS
28. Are buildings and equipment in good repair and clean?
29. Are only approved maintenance and cleaning chemicals used, and are they properly identified, handled, and stored?
30. Are animals and vermin excluded from the area?
31. Are food contact surfaces properly cleaned to prevent contamination?
32. Are utensils and portable equipment stored in clean locations?

EQUIPMENT AND PROCEDURES
33. Is equipment of proper design and sanitary construction?
34. Are all materials and working fluids free of PCB's?

TABLE 5.1 (*continued*)
PLANT INSPECTION CHECKLIST FOR GMP'S*

PROCESSES AND CONTROLS

35. Do all operations in the receiving, transporting, packaging, preparing, processing, and storing of food follow sanitary principles?
36. Are raw materials and ingredients properly inspected and segregated from processed and finished products?
37. Are containers and carriers of raw materials inspected?
38. If ice is used in contact with food, is it sanitary?
39. Are food processing facilities dedicated to food products?
40. Is processing equipment routinely cleaned and inspected?
41. Is proper control maintained over times, temperatures, humidity, pressure, and other process variables to minimize spoilage and microbial contamination?
42. Are proper testing procedures used to check the quality of ingredients and finished products?
43. Are approved packaging materials used that will provide suitable protection to the products?
44. Are products properly coded and are records retained?
45. Are products shipped and warehoused under sanitary conditions and away from harmful substances?

*Violations may be classified by the following ratings to indicate their seriousness:

Class A is a minor infraction that should be corrected on a routine schedule.

Class B is a serious condition that requires prompt attention.

Class C is a critical defect that can directly lead to the production of adulterated or misbranded product. It must be corrected immediately.

Class R is a repeat item which is indicative of a poor or slow response.

MEAT, POULTRY, AND EGG PRODUCTS

USDA administers regulations similar in scope to the GMP's promulgated by FDA. The Food Safety & Inspection Service (FSIS), formerly known as the Food Safety & Quality Service (FSQS), is responsible for the inspection of meat, poultry, and egg product plants and the enforcement of the respective sanitary standards established for these products (9CFR308, 9CFR381.45, 7CFR59.500). In addition, the Agricultural Marketing Service (AMS) of USDA is charged with the grading and inspection of dairy plants for compliance with approved specifications (7CFR58).

USDA Plant Inspections

Until 1986 the authorizing statutes required that USDA maintain around-the-clock inspection over the plants under its jurisdiction. In practice, however, FSQS Administrator Robert Angelotti acknowledged in 1978 that "The provision is not now, and probably never has been, literally applied.'' He noted that USDA maintained what it called ''continuous supervision'' rather than ''continuous inspection'' over these plants. Furthermore, he disclosed that half of the federally inspected meat and poultry plants were visited by FSQS inspectors on a ''patrol'' basis, equivalent to as little as one hour a day.[19]

Heralded as the most significant change in the meat inspection statute during the past 81 years, an amendment was passed by Congress in October, 1986, that eliminated mandatory continuous inspection. At the same time, the bill strengthened the enforcement powers of USDA by giving the department authority to shut down temporarily a meat packer in flagrant violation of the law. Under reduced inspection plans, the department expected that qualified plants would be inspected no more than twice a week and in some cases only once a month. Furthermore, a department spokesman predicted that the need for prior approval of all labels would be discontinued.[20,21,22,23]

Besides products shipped in interstate commerce, USDA has assumed responsibility for all meat inspection activities of the Department of Defense.[24] At the request of many states, USDA has also taken over the inspection of plants supplying meat and poultry products for intrastate commerce. In 1978 New Hampshire became the eighteenth state relying on USDA for meat inspection and twenty-sixth state for poultry inspection.[25] All meat plants coming under USDA supervision must meet the guidelines in Agriculture Handbook 191 although some allowances may be made during a transition period.[26] Irrespective of quality considerations, only USDA-inspected products can be shipped interstate or exported.

Total Quality Control Program

Under the inspection policy established by USDA, manufacturers too often have come to rely on inspectors to detect improper conditions rather than take the initiative in correcting problems. The weakness in the system plus the growing demands placed on the limited staff of inspectors caused USDA to reassess its procedures in the late 1970's. Looking for alternatives, the department turned to FDA's now defunct CQAP from which it borrowed several concepts while rejecting other aspects. After testing its proposals in a pilot program, USDA published on August 15, 1980, in the *Federal Register* a voluntary Total Quality Control (TQC) program. The important features of the program are as follows:

• A plant wishing to join TQC must submit to USDA a plan which specifies the critical control points of its quality assurance program. All proprietary data will be protected from disclosure under the Freedom of Information Act (FOIA).[27]

• An approved plant would remain under ''continuous inspection'' by USDA, but savings could be anticipated in reduced overtime compensation for USDA inspectors.

• Formal reports are kept to a minimum. Plants in the TQC program are not required to submit routine exception reports to a central reviewing office inasmuch as supervision is maintained by USDA field personnel.[28]

• Small plants are encouraged to participate in the program. As an aid to these processors, USDA has prepared a booklet titled *Small Plant QC Guidebook*.

• Allowances will be made under the program for the development of new products.

• A uniform logo may be used by an approved plant to identify its products. The seal shows an arrow with a feedback loop enclosed within a diamond border. The words "Quality Control USDA Approved" are inscribed on the logo.[29]

• As a measure of the success of TQC, 205 plants had been approved by FSIS as of September, 1983. Of this total, 173 plants were operating under the program.

SANITATION AND PEST CONTROL

The most tangible evidence of good manufacturing practice is sanitation. Therefore it is not surprising that sanitation was the first food-related issue to catch the public's attention when in 1906 Upton Sinclair published his novel, *The Jungle*, about the prevailing conditions in meat packing houses. One of the main characters of the book recounted his third day on the job as follows:

> It seemed that he was working in the room where the men prepared the beef for canning, and the beef had lain in vats full of chemicals, and men with great forks speared it out and dumped it into trucks, to be taken to the cooking-room. When they had speared out all they could reach, they emptied the vat on the floor, and then with shovels scraped up the balance and dumped it into the truck. This floor was filthy, yet they set Antanas with his mop slopping the "pickle" into a hole that connected with a sink, where it was caught and used over again forever; and if that were not enough, there was a trap in the pipe, where all the scraps of meat and odds and ends of refuse were caught, and every few days it was the old man's task to clean these out, and shovel their contents into one of the trucks with the rest of the meat![30]

Although remote in time, this story illustrates lessons that continually need to be relearned. In 1984, Colorado-based Cattle King Packing Co. was convicted of selling adulterated meat from its packing plant in the Denver area. According to testimony at the trial, the facility was infested with rats and roaches while paint chips littered the plant. Diseased carcasses were dragged into the slaughter house and ended up as rotten meat in hamburger. Witnesses described people urinating on the floors of the packing plant. These accounts were all the more shocking in the light that Cattle King was the largest supplier in the nation of ground beef to school-lunch programs.[31]

Sanitary Materials of Construction

Upton Sinclair's vivid portrayal of the meat industry convinced people of the need for sanitation, but before headway could be made, the right food processing equipment had to be developed. This effort began to gather momentum during the 1920's when extensive studies were made to determine the proper materials for use in food contact surfaces. Up to this time, such metals as iron, steel, brass, bronze, copper, tin, and galvanized iron had been used in spite of their unsatisfactory performances. They all corroded badly, especially in the presence of food acids. They produced off-flavors, and none of them could be properly cleaned. Only pure nickel gave satisfactory results, but its prohibitive cost restricted its application.

Manufacturers of equipment for the dairy industry led the search for a sanitary metal. After exhaustive experiments with over 400 alloys, Loomis Burrell of D. H. Burrell & Co. discovered in the late 1920's a composition of nickel, copper, and other metals which performed satisfactorily. Given the trade name, "Diamond" metal, it was first used in castings for sanitary milk pipeline fittings. Simultaneously with Burrell's development, the Waukesha Foundry Co. came up with German Silver, a cupro-nickel alloy which the firm called "Waukesha Metal." It was used for casting pump parts, fittings, and valves. Both of the alloys developed by Burrell and Waukesha are known as "dairy" or "white" metal. They can easily be worked, readily cleaned, and to this day they provide excellent service.

Across the Atlantic, Fried. Krupp in Essen, Germany, discovered stainless steel in 1908. A ferroalloy containing approximately 18 percent chrome and 8 percent nickel, this material was found to possess outstanding properties but required completely new techniques for its fabrication. After World War I Allegheny Steel took the lead in developing stainless steel sheet, strip, and bars in the United States. These products caught the attention of the J.G. Cherry Co., which made the first pasteurizer out of this material in 1926. Two years later the Cherry and Burrell companies merged to form the Cherry-Burrell Corporation.[32]

Equipment Design and Standards

Of significance equal to the development of materials was the establishment of design specifications for food processing equipment. The proliferation of sizes, threads, and designs prevented parts from different manufacturers from being used interchangeably. In the early 1920's, representatives from the International Association of Milk Dealers (which became the Milk Industry Foundation) and the Dairy and Ice Cream Machinery & Supply Association (now the Dairy and Food Industries Supply Association) met to simplify the specifications for pipe, fittings, and outlets on processing equipment. This group was joined by a committee from the International Association of Milk and Food Sanitarians. Because the original standards were established by three cooperating associations representing dairy processors, equipment manufacturers, and sanitarians, the code came to be known as 3-A Sanitary Standards. Participation has been increased by replacing the founding organizations with the International Association of Milk, Food, and Environmental Sanitarians; the U.S. Public Health Service; and the Dairy Industry Committee. The latter group represents the American Butter Institute, American Dry Milk Institute, Dairy and Food Industries Supply Association, Evaporated Milk Association, International Association of Ice Cream Manufacturers, Milk Industry Foundation, National Cheese Institute, and Whey Products Institute.

Following the pioneering work of the dairy industry, other food industry segments have established their own equipment standards. As an outgrowth of the 3-A Sanitary Standards, the E-3-A Standards were developed for egg processing equipment. In 1949, six bakery associations formed the Baking Industry Sanitation Standards Committee (BISSC) to attack serious sanitation problems caused by poorly designed equipment. The Association of Food and Drug Officials published in June, 1961, the AFDOUS Frozen Food Code which includes specifications for processing equipment. USDA issues a quarterly bulletin, *Accepted Meat and Poultry Equipment*, which lists approved equipment for all plants inspected by this agency.[33] Meanwhile Underwriters Laboratories has taken the initiative to establish safety standards for industrial as well as consumer products. Incorporating many of the above standards, the American Society of Mechanical Engineers (ASME) and the American National Standards Institute (ANSI) have developed Sanitation and Safety Standards for Food, Drug, and Beverage Equipment, available as report ANSI-ASME F2.1-1975 and its addendum F2.1A-1976.

The voluntary standards developed by and for the food industry have been lauded as outstanding achievements in cooperation. Government has participated in much of these efforts to the benefit of all concerned. In spite of the generally favorable reaction to these initiatives, the Office of Management and Budget (OMB) issued in January, 1980, a report, Circular A119, that was critical of this activity. Calling for the establishment of a uniform policy in the operation of standard-setting groups, OMB recommended certain "due process" procedures requiring public notices, open hearings and detailed records. Questions have also been raised concerning potential anti-trust violations as a result of issuing standards. (Similar concerns hold for establishing voluntary product grades.) Whatever the direct costs to manufacturers, most observers believe that independent organizations are far better qualified to provide leadership in standard-setting than is the federal government.[34]

While the original purpose for setting equipment standards was greater uniformity between manufacturers and a reduction in the number of sizes, the eventual rewards were of far greater consequence. Out of these considerable efforts, some very significant principles were formulated for sanitary equipment design. A summary of some of the more important points to consider is presented in Table 5.2. These specifics give meaning to the requirement under the GMP's that equipment be of proper design and sanitary construction.

TABLE 5.2
PRINCIPLES OF SANITARY EQUIPMENT DESIGN[a,b,c]

1. All food contact surfaces shall be inert, wear resistant, and smooth. Pitting and crevices must be avoided.
2. No food contact surface should be painted.
3. Copper and copper alloys should not be used in contact with edible oils and other fatty foods.
4. There should not be any contact between dissimilar metals in food service.
5. Permanently joined metal surfaces exposed to food should be welded rather than riveted. The seams should be ground and finished flush.
6. There should not be any screws, rivets, or bolts projecting from food contact surfaces.
7. Excessive mechanical stresses should be avoided.
8. The radius of all inside corners should be 1/4 inch or greater for easy cleaning.
9. Localized hot spots should be avoided in heat transfer equipment.
10. All surfaces in contact with food should be accessible for inspection.
11. Hatch openings in vessels should be raised 1 to 6 inches above the outside surface, and vessels should be installed so that openings are above the level of the surrounding deck or floor.
12. Equipment, piping, and valves are required to be self-draining.
13. All unnecessary piping should be eliminated, and dead ends and pockets must be avoided.
14. Sharp bends in piping should be avoided.
15. Equipment should be supported by structured tubing sealed at the ends, not by angle irons.
16. Piping is correctly suspended by using trapeze or clevis type hangers, not chains or threaded rods.
17. Seals and gaskets in contact with food shall be non-toxic, non-absorbent, non-exuding, and inert.
18. Whenever possible it is preferable to transport dry bulk materials by fluidization rather than by screw conveyors or bucket elevators.
19. Covers should be installed over tanks, hoppers, flumes, conveyors, and other open equipment to prevent debris from entering the system.
20. Open grating for catwalks and stairs should not be used over processing areas.
21. Seals and bearings of rotating equipment need to be externally located to prevent lubricant from leaking into the product.
22. Avoid toxic fluids in sensing devices.
23. The air supply shall be properly filtered or washed, dried, and free of oil.

[a]Wm. S. Stinson, "Sanitary Design Principles for Food Processing Plants," *Food Processing*, Mid-July, 1978, pp. 98-108.

[b]Robert Bannar, "Safe, Sanitary Equipment: Do You Have It?" *Food Engineering*, March, 1979, pp. 111-113.

[c]Lyle D. Perrigo, "Design to Reduce Corrosion," *Food Technology*, January, 1975, p. 54.

Cleaning of Equipment

How well any food processing equipment has been designed and built will determine the ease with which it can be cleaned. Proper cleaning is equally dependent on an expert evaluation of the job to be done and the correct selection of cleaning compounds and conditions. Generally the cleaning procedure will consist of six sequential steps: pre-rinse, clean, intermediate rinse, sanitize, post-rinse, dry. One or more of these steps may be omitted or combined with another step in special circumstances. For example, surfaces sanitized with an iodophor can be dried spot-free without using a post-rinse. The drying step can be eliminated provided the surfaces do not come into contact with dry product. In some instances the cleaning and sanitizing steps can be combined by using such specially formulated products as those containing anionic surfactants and acid. By eliminating steps in the cleaning cycle, substantial savings may be realized by reducing downtime, water requirements, energy consumption, and the usage of chemicals.

Effective detergent formulations may contain several additives including a surface-active agent (surfactant), a chelating agent (sequestrant), and either an alkali or acid. The choice of additives will depend on the type of soil to be removed from the equipment, the materials of construction, and the method of applying the cleaning compound. The surfactant, commonly either anionic or non-ionic, promotes rapid wetting, penetration, and the emulsification of fats and oils. It also aids in the dispersion and suspension of dirt particles. Inorganic alkalis including caustic soda, sodium metasilicate, and trisodium phosphate are effective in removing fats and protein and such difficult to dissolve deposits as tars formed in smokehouses. To tie up calcium ions in alkaline solutions, sequestering agents, e.g., polyphosphates, gluconic acid, or ethylenediamine tetracetic acid (EDTA) are critical.

To clean equipment that is subjected to elevated temperatures during food processing, an acid formulation would be selected. Containing phosphoric, nitric, or sulphamic acid, such a cleaning compound is capable of dissolving carbonate scales and certain mineral deposits like milkstone, eggstone, and beerstone. An acid wash may be preceded or followed by an alkaline cleaning step. Frequently, as in dairy plants, an alkaline wash is used first and then an acidic sanitizing solution is applied to brighten the equipment by neutralizing excess alkalinity and preventing the buildup of mineral deposits. Most importantly the mild acid solution passivates the stainless steel, making it inert to corrosive attack.[35,36]

Sanitizing Compounds

Once the equipment has been cleaned, it must be sanitized to destroy residual yeast, mold, bacteria, and spores. Experts repeatedly warn that any attempt to sanitize surfaces that are not absolutely clean is done so in vain. Although a conscious effort is made to approach total kill of all organisms, no pretense is made that the equipment will be sterilized or disinfected. Such extreme action would be impractical and unnecessary in an environment that is permeated with microorganisms. One way to achieve sanitation is by the application of heat, e.g., steam, but this method is difficult to control. Therefore the preferred approach is to employ bactericidal chemicals.[37]

To be judged an effective sanitizer, a compound must pass an efficiency test which requires that 99.999 percent of the harmful organism be killed within 30 seconds. Three standard methods may be used for determining the overall effectiveness of a sanitation procedure: the use of rinse solutions, swab procedures, or replicate organism direct agar contact (RODAC). While these test methods are officially accepted by regulatory agencies, they have the disadvantages of being laborious, time consuming, and requiring experienced technicians. For these reasons a more efficient method involving adhesive strips is under development. This procedure, however, still requires the same incubation time and therefore falls short of qualifying as a truly rapid method.[38]

FDA has passed on the acceptability of some twenty-two sanitizing solutions (21CFR178.1010), but there is some uncertainty about the completeness of this list. Some solutions may not be included that were approved by the Public Health Service before 1958. Other products may be covered by prior sanction. EPA reportedly has advised that any of the approved solutions can be mixed with chemicals that are generally recognized as safe. To dispel any possible confusion, FDA issued guidelines in 1986 for food additive petitions covering sanitizing solutions.[39] The following four types of approved sanitizers, however, fulfill most of the needs of food processors.

Chlorine compounds, typified by hypochlorites, are the most economical of the common sanitizers and thus most widely used. With excellent germicidal power, they are effective against all microorganisms, bacteriophage, and even spores if the temperature is sufficiently high. They are relatively non-toxic at use strength of less than 200 ppm chlorine, and they do not form films. Chlorine compounds do have several drawbacks. They have limited shelf life and are corrosive to most metals.

Iodophors, which are combinations of iodine and solubilizing agents, possess good stability. They are active against all microorganisms except spores and bacteriophage. Generally used at a concentration of 25 ppm iodine and under acidic conditions, iodophors exhibit good penetration and do not leave a film on drying. They are non-corrosive and non-irritating to skin. Their amber color provides an indication of the presence of active iodine. Principal disadvantages are that they are relatively expensive and are limited to temperatures under 120°F. (49°C.)

Quaternary ammonium compounds, or quats as they are commonly called, provide better control of gram-positive bacteria including staphylococci than the gram-negative bacteria, such as coliforms and psychrophiles, e.g., pseudomonas. They are ineffective against spores and bacteriophage. They have a long shelf life and are non-corrosive and are therefore suitable at higher temperatures than permitted with hypochlorite sanitizers. At the recommended concentration of 200 ppm, quats possess considerable detergency and provide excellent penetration; however, with mechanical agitation they may cause foaming. Negative aspects include their higher cost and incompatibility with anionic surfactants.

Acid-anionic surfactants are effective only at a lower pH, the optimum range being 1.9 to 2.2. Used at 100 ppm of ionic surfactant, they are capable of controlling a wide spectrum of microorganisms, including some thermodurics, but they are ineffective against spores. These sanitizers are stable, non-corrosive to stainless steel, and can be used at higher temperatures. They are effective in removing such mineral deposits as milkstone. Their chief disadvantages are that they are corrosive to metals other than stainless steel, and they present foaming problems in mechanical systems.[40,41]

In the selection of cleaning compounds and sanitizers, care must be taken to prevent corrosion of the food contact surfaces. In bakeries, for example, pans and utensils may be made of softer metals like aluminum and tin, which cannot tolerate strongly acidic or alkaline solutions. In these applications buffered compounds with inhibitors should be used.[42] Neither is stainless steel immune to corrosion. Hypochlorite solutions must be kept alkaline and used at moderate temperatures and concentrations. Additionally, care needs to be taken to avoid the presence of chloride ions under oxidizing conditions and elevated temperatures, such as might be experienced in an evaporator. If these conditions cannot be avoided, either titanium or high nickel alloys, such as Hastelloy C, should be used. One should also keep in mind the corrosive properties of sulfuric acid which even in dilute solution will attack stainless steel. Pitting and crevice formation on food contact surfaces caused by corrosion are extremely detrimental to proper sanitation.[43]

Application of Cleaning/Sanitizing Solutions

Several methods of application can be used for cleaning and sanitizing solutions. Before the introduction of automated procedures, manual methods relied on brushes and plenty of "elbow grease." Next came steam and steam-water mixes, which at the time were considered state-of-the-art. For the most part these methods have been supplanted by chemical systems. Foam applications, which increase the contact time between the cleaning compound and the surface, are used for floors, walls, and machinery. Removable parts and equipment such as kettles, pans, and meat racks generally are soaked or cleaned-out-of-place (COP).

The newest and most efficient method for cleaning closed equipment is clean-in-place (CIP). This approach entails either the circulation of solutions through the processing lines and equipment or the application of jets through nozzles and pressure spray balls.[44]

A CIP system must be meticulously engineered to guarantee positive results. As one professional stated the case, "97% clean is still 3% dirty."[45] The last remaining 3 percent uncleanliness can spell the difference between producing acceptable product and generating filth. Successful cleaning is achieved by the careful adjustment of the following variables:

Time of contact with the cleaning solution is important to allow penetration of the soil film and its removal. The suggested length of time ranges from 10 to 20 minutes for cleaning cold surfaces and 15 to 30 minutes for equipment in hot service. Excessive times will only lead to lost production and an unwanted drop-off in the temperature of the cleaning solution.

Time is also a factor in the application of sanitizing solutions. Studies have shown a logarithmic relationship between the number of microorganisms killed and the time of exposure.

Temperature of the cleaning solution is typically held around 160° to 185°F. (71° to 85°C.) for cleaning hot surfaces and somewhat lower, 135° to 160°F. (57° to 71°C.), for cold surfaces. As a rule each 20°F. (11°C.) increment in temperature will double the activity of the cleaning agent. Higher temperatures also increase the rate of kill by sanitizers. Upper limits are set by consideration of sanitizer stability and corrosion rates.

Concentrations of the cleaning solutions for optimum results have been determined by the suppliers and are indicated on the labels. Most alkaline cleaners work best at around 0.5 to 1 percent by weight. Acid cleaners are usually adjusted to a desired pH range. For sanitizers, increasing the concentration accelerates the destruction of bacteria. Maximum use levels are specified by FDA for approved sanitizing solutions.

Physical action is of utmost concern in cleaning. To obtain the necessary agitation, a velocity of 5 ft. per sec. or greater is required. Compensations must be made if different sizes of pipes are installed in the same line. Sufficient pressures must be supplied to operate spray devices which have to be correctly designed and placed to ensure complete irrigation of all surfaces. A new development is the application of bursts of spray for 20 to 45 second durations.

CIP systems can range from a manual hookup to a semi-automatic operation or one that is fully controlled by a microprocessor. The decision concerning which configuration to use will depend on the complexity of the facility and the potential savings in cleaning time, materials, and utilities. In designing a CIP layout, as many items of equipment as possible should be grouped together in each cleaning circuit. Usually a large plant is divided into several such circuits, some being cleaned while others are in production. To place a circuit in the cleaning cycle, the operator may have to break into the system to connect the lines to the supply of cleaning solution. Accidental intermixing of product and detergent can be prevented by designing the plant with "key pieces" of pipe sections that fit between only two points in the system. Another precaution to take is to insert air vents on tanks and vessels to avoid collapsing them when a vacuum is pulled. A well designed CIP system will optimize investment and operating costs.[46]

To illustrate one area where substantial savings are possible in CIP, it is instructive to take a close look at the water requirements. Water management will not only yield direct benefits but can save on energy, materials, and sewer demand. The determining factor in conservation is water reuse. By using the principle of counter-current flow, large water savings can be realized without in the least compromising sanitation. Instead of wastefully discharging the post-rinse to sewer, it frequently can be recycled to the pre-rinse stage. Make-up water for the cleaning solution can be taken from the intermediate rinse. These procedures, when combined with other conservation measures throughout the plant, can produce substantial savings.[47,48]

Pest Control

A total program in sanitation includes control over all pests: rodents, flying insects, crawling insects, and birds, all of which are carriers of disease and filth. Ranking these vermin in the order of greatest nuisance, rodents have been likened to "the most (not) wanted public enemies of the food industry."[49] FDA has reported that rodents are responsible for 90 percent of the complaints filed against food establishments, and mice infestations account for 90 percent of all rodent complaints. Unfortunately there are no pat answers to controlling rodents or any of the other pests. A combination of partial solutions is needed to maintain a dynamic balance. Their populations can be kept at a minimum by (a) eliminating their breeding places and natural habitats in the plant environs, (b) restricting their access to food storage areas, and (c) setting up extermination procedures to eliminate those pests which eventually find their way into restricted locations.

To keep rodents from entering a plant or warehouse, close scrutiny of all incoming goods should be maintained. Carriers are notorious offenders in contributing to infestation problems. Additional precautions should be taken by making the building as tight as possible. All openings should be closed or screened, and cracks must be caulked. To control rodents inside the facility, poison baits containing anticoagulant are effective in killing rats but not mice. The latter's feeding habits, characterized by nibbling, generally are not conducive to their taking enough bait to harm themselves. The best method of catching mice is by setting traps. Ultrasonic devices have been promoted for rodent control but are still experimental. Although the generated sound is a high pitch and therefore inaudible to human ears, there have been some concerns expressed about the possible physiological effects on people.

Flying insects include houseflies and drosophila gnats or fruit flies. A single innocent-looking housefly can carry an estimated 3,680,000 bacteria.[50] To prevent flying insects from coming into a plant through doorways and entrances, air curtains have been found to be successful although not foolproof. Such devices require a minimum air velocity of 1600 ft. per minute (close to 18 mph), and care should be taken to maintain an even or positive atmospheric pressure within the plant.[51] Drosophila usually enter a plant with overripe and rotten fruit. To control flying insects inside a plant, one can resort to electric grids and space spraying or fogging with approved insecticides such as pyrethrum.

Birds and bats are more troublesome than they might seem. Starlings, sparrows, and pigeons will find their way into a plant or warehouse through doors or any unscreened openings in the building. To get rid of birds, the best advice given is to ''pester the pest.''[52] The first step is to eliminate nesting and roosting sites under the eaves and among the rafters. Porcupine-like barriers made of stainless steel quills have been placed on ledges to keep birds from lighting. An imaginative array of devices are offered to deter birds from remaining in a building. A revolving amber light is highly irritating to their eyes. A chemical repellent sold under the trade name, Bird Trip, deprives a bird of its senses and upsets its routine behavior. If all else fails, the suggestion has been made, only half in jest, to obtain a fake owl to frighten the birds away.

Crawling insects, including beetles, mites, silverfish, cockroaches, and moths are a major menace. The first rule in controlling these pests is to maintain cleanliness. One must begin with the stock rotation of all inventories, including raw materials, packaging supplies, and finished product. Spills should be promptly swept up or vacuumed, and ripped bags and broken containers set aside. All sources of water, such as leaks in the roof, should be eliminated. Roaches and other pests look for places to hide. Therefore, materials should be neatly stacked in aisles or rows and set back 18 inches from the walls. This arrangement also makes cleaning and inspection easier. Bags and drums should be placed off the floor and on pallets. Regardless how good a job is done on housekeeping, there will be a need for fumigation.

The recent curtailment of ethylene dibromide (EDB) severely restricts the choice of fumigants. The application of chemical fumigants is regulated by FDA and the Environmental Protection Agency (EPA). Under the Federal Insecticide, Fungicide, and Rodenticide Act (FIFRA), EPA is entrusted with the task of overseeing pesticide manufacture and usage. At the same time FDA is responsible to make sure that food is not adulterated by any pesticides.[53] Certain pesticides are classified as restricted for use by or under the direct supervision of certified applicators. These persons must attend training programs which are administered by the states. Plants which do not have their licensed applicator in-house must contract with a commercial applicator for his services.[54]

A Case History of Poor Sanitation

Sanitation requires constant vigilance. When management relaxes its controls over operations, it must be prepared to face the same humiliation experienced by the Good Humor Corporation in 1975. This company, whose name is a household word, was indicted on charges of selling adulterated product from its Maspeth, Queens plant in New York. Records were allegedly falsified, and documents were shredded to cover up evidence that batches of ice cream contained excessive levels of coliform bacteria "TNTC" (Too Numerous To Count.) State health officials blamed slipshod sanitation for the results. The company pleaded innocent of any wrongdoing, but when the press was finished with its coverage of the episode, consumers were muttering about "Bad Humor."[55,56]

EMPLOYEE TRAINING

Without adequate training, personnel have proven to be the weak link in quality assurance. All too frequently human error is cited as the cause of product failure. Motivated by a desire to improve productivity, corporations are investing heavily in automation. As more and more employees are being replaced by computers, robots, and instruments, it is fair to ask what can people do better than machines. The answer is simple: only people can be taught to think — to observe unexpected results and to deduce the significance of their findings. The greatest challenge in any training program therefore is to instill a sense of inquisitiveness in the personnel responsible for quality assurance.

Development of Skills

For the sake of job training, each function can be broken down into a blend of technical and administrative skills. The necessary level of technical competence will depend on the position to be filled. The line worker, for instance a laboratory technician, will require proficiency in routine tasks, such as protein analysis, the supervisor must possess professional expertise, and supporting personnel should have a familiarity with the relevant technology. Similarly, administrative skills should match the needs of the assignment. There may be varying requirements to communicate results, make financial decisions, and manage projects. The development of technical and managerial aptitudes begins in school or college and extends throughout the career of the employee.[57]

Many helpful training materials are available from government and management consultants. Films, slides, brochures, and posters illustrate important principles in a memorable fashion that will catch the attention of employees. As provided by FDA's "Food Industry Education Programs," help can be sought in three areas, Industry Information Materials and Assistance, Industry Quality Assurance Assistance Program (IQAAP), and Foreign Government Assistance

Program.[58] The Industry Program's Branch (IPB) of the Bureau of Foods working through field offices stands ready to respond to the requests for information and assistance from manufacturers, food industry segments, and importers of food products. (FDA on March 19, 1984, reorganized its Bureau of Foods and changed the name to the Center for Food Safety and Applied Nutrition. The reorganization established five new offices: Office of Management, Office of Compliance, Office of Toxicological Sciences, Office of Physical Sciences, and Office of Nutrition and Food Sciences. In turn these offices were subdivided into seventeen divisions.)[59] To assist the public in ordering reprints, FDA has issued a pamphlet describing its manuals.[60] In addition, *FDA's Catalog of Information Materials for the Food and Cosmetic Industries* lists available publications, audiovisual aids, and copies of laws and regulations that are available at nominal charges.[61] One of the outstanding presentations is a movie called *Purely Coincidental* which tracks the parallel performances of two fictitious but realistic operations, one a food plant and the other a pet food establishment. From time to time FDA has offered instructional booklets for circulation to workers and managers. Recent titles include *So You Work in a Food Plant!*[62] and *Follow the Signs to Safe Food.*[63]

For assistance in training its employees, organizations can turn to several industry associations which schedule workshops and sponsor seminars. The American Society of Quality Control (ASQC), the Food Processors Institute (FPI), Grocery Manufacturers of America (GMA), and the Institute of Food Technologists (IFT) in the past have all conducted short courses on topics of timely interest. Formal training programs may also be required by the regulations. As noted in the discussion of sanitation in this chapter, EPA requires the training and certification of persons applying pesticides. FDA specifies that the operation of retorts in the production of low-acid canned foods must be under the supervision of a person who has attended an approved school in this technology. (21CFR113.10)

The umbrella GMP's provide that all food handlers and supervisors receive appropriate training. (21CFR110.10) In planning a training session, the agenda should be prepared keeping in mind that employee participation greatly enhances their retention of the subject matter. Accepted as a successful teaching tool in the leading business schools, case studies are effective in stimulating discussions on important points. A record should be kept of each meeting indicating the topics reviewed and the persons in attendance.

Personal Hygiene

One topic of overriding concern for food workers is personal hygiene. People account for a major source of microbiological contamination and are the most difficult variable to control. Operators continually shed dust particles and microbes from their skin, noses, mouths, and clothing. Studies have demonstrated that each individual can contribute thousands of airborne bacteria every minute. One cough or sneeze can push this number into the millions. This hazard has long been recognized in hospitals, especially in surgical operating rooms. More recently, ultra-clean rooms have been installed by manufacturers of drugs, spacecraft, and electronic equipment. While the food industry does not have to go to these extremes, it must discipline its workers to comply with the few but necessary precautions outlined in the GMP's.[64]

Those managers who have been responsible for quality assurance repeatedly testify that objectives cannot be attained by proclamation alone. "For it is people who actually set the rules, follow the rules, and also break the rules."[65] The importance of worker attitude was stressed in covering quality circles in Chapter One. Persuasion, not imposition, should be the watchword in quality assurance. And effective persuasion requires education.

GOOD LABORATORY PRACTICE

The impetus for promulgating Good Laboratory Practice regulations (GLP's) was a scandal that erupted in 1976 over the registration of a new drug. Industrial Bio-Test, Inc. (IBT), an independent testing laboratory, was caught and later convicted of falsifying data submitted to FDA. Founded in the 1950's, IBT for years was one of the fastest growing laboratories in the country. Outsiders credited the firm's popularity to the fact that it consistently produced the "right" test data — those results that would quickly get a product past the safety and health review processes of FDA and the Environmental Protection Agency (EPA). The disclosure of IBT's fraud brought about the quick demise of the business in 1977, but subsequent investigations of other laboratories revealed many instances of irregularities and sloppiness in the testing of new products.[66]

The Senate Subcommittee on Health and Scientific Research held a series of well publicized hearings in 1978 to look into the issue of laboratory practices. Responding to public pressure, both FDA and EPA proposed regulations for overseeing this activity. FDA finalized GLP's on December 22, 1978, for all nonclinical investigations of drugs and food additives. These studies include in vitro experiments and in vivo animal tests designed to generate toxicity data. Basic research is not affected by the regulation, but foreign laboratories that provide supporting evidence for U.S. registrations must abide by the provisions. FDA is free to reject any data from a laboratory that does not adhere to the GLP's, and in more serious infractions the agency is empowered to disqualify a laboratory as a testing facility.[67]

The GLP's do not require the development of more extensive data; they simply specify formal directives for obtaining the necessary information. Before a study is undertaken, a protocol for the test must be agreed upon, and this plan must be followed barring unforeseen developments. It is incumbent on the laboratory to show that all personnel are qualified by their education and training to conduct the experiment. Test substances must be completely defined as to purity, quality, and strength. Facilities and equipment must be appropriate to the needs of the laboratory, and they should be well maintained. Considerable emphasis is placed on the correct reporting of test data and the proper documentation of these results. Finally, the GLP's provide for a Quality Assurance Unit (QAU) made up of one or more individuals who, acting independently from the personnel engaged in the laboratory work, can vouch for the validity of the data.[68]

INFANT FORMULA

The medical profession uniformly endorses breast feeding, but it recognizes that for physiological, sociological, or other reasons this choice may not be available. Before the twentieth century the only recourse was to procure a wet nurse if one could be found. Today the accepted alternative is to put the baby on infant formula. These products have been perfected to the point where their nutritional profile closely resembles mother's milk. Convenience rather than need is now the primary reason why many mothers in industrialized nations decide on infant formula.

Infant Formula Industry

The foundation of the infant formula industry was laid by Gail Borden, who received a patent in 1856 for his process to produce condensed milk. Later he learned how to sterilize the product, thus allowing its wide distribution. The process modified the casein in the milk, rendering it digestible to infants. On the other hand the heat treatment had a negative effect by destroying much of the anti-scorbutic value of the milk, necessitating supplementation with vitamin C. Until the 1950's condensed milk mixed at home with corn syrup was the mainstay of infant formulas.

The first commercial infant formula was introduced in 1915. It contained various vegetable and animal fats and oils homogenized with skim milk. Later soy protein was substituted for milk in certain formulas for babies allergic to cow's milk. Since 1960 commercial products have virtually monopolized the infant formula market, and according to an FDA report, by 1982 annual sales of these products had risen to over $500 million. At one time half of all infant formulas were in powder form, which was originally developed for consumption in hospitals. Today most products are sold in liquid forms, either ready-to-feed or as a concentrate which first needs to be diluted.[69]

Safety Record Broken

The safety record of the infant formula industry for years was regarded as impeccable until a mishap occurred that had a lasting impact on the business.

In July, 1979, physicians around the country reported that babies on two soy-based formulas, Neo-Mull-Soy and Cho-Free, were suffering symptoms from a condition called metabolic alkalosis. The infants had lost their appetites and were failing to thrive or gain weight. The manufacturer, Syntex Laboratories, was notified, and on checking retained samples, it found that the chloride content was below acceptable levels. The company had reduced the salt content of its formulas on purpose, but because it was not doing a complete analysis of its finished products, it failed to catch the chloride deficiency. Close to 140 babies were afflicted. While most of the infants later recovered from the illness, at least one child is alleged to have developed speech and memory disorders.[70]

Infant Formula Act

The public reacted with indignation to the Syntex incident. This concern was reflected in Congress, which acted swiftly to establish new codes of conduct for the industry. On September 26, 1980, President Carter signed the Infant Formula Act of 1980, which strengthened FDA's surveillance of these products. The new law, incorporated into the FD&C Act as Section 412, provides that:

- Manufacturers of infant formula must comply with quality control regulations prescribed by FDA for these products.

- Notification prior to manufacture must be given to FDA of any new product or changes in the formulation or processing of existing products.

- All health hazards and product deviations must be immediately reported to FDA.

- Product recalls shall be carried out in accordance with FDA regulations.

- Distribution records must be maintained for a period up to two years.

- With the exception of products designed for special needs, infant formulas must contain the nutrients set forth in the included table.

Questions surrounding the clinical testing of infant formulas were left open. As a matter of course, manufacturers have assumed much of this responsibility. Support also has been provided by the American Academy of Pediatrics (AAP), which has submitted its recommendations for the nutrient content of these products.

Quality Control Regulations

Substantive revisions in manufacturing practices had to await the publication of FDA's quality control regulations. After considerable debate on proposed regulations, FDA issued on April 20, 1982, final rules which provided:

- In general, a manufacturer is required to sample and analyze each batch of infant formula for most required nutrients. A manufacturer may do so either by sampling and analyzing each batch of finished product or by sampling and analyzing ingredients and conducting in-process testing on each batch.

- Each manufacturer is further required to conduct stability testing of representative samples over an infant formula product's shelf life to confirm maintenance of nutrient content.

- Each manufacturer is required to conduct additional tests, which are not required in normal production, on new formulations and after major processing or formulation changes.

- The final rule further requires each manufacturer to code all infant formula containers and to maintain and make available to FDA investigators quality control records.[71]

As noted in the above quality control provisions for infant formula, a manufacturer could elect to establish a quality control system based on "sampling and analyzing ingredients and conducting in-process testing on each batch." Instead of analyzing for each nutrient, FDA proposed that for control purposes an "indicator nutrient" should be selected from each premix of known composition. By following the concentration of the indicator nutrient(s) throughout the process, the manufacturer could then keep tabs on the composition and uniformity of his product. If the chosen indicator nutrient was the one that was most susceptible to process damage, e.g., from heat, light, oxygen, or pH, it could also be used to measure product degradation. As examples, vitamin A was suggested as an indicator nutrient for an oil-soluble premix, vitamin C for a water-soluble premix, and manganese for a mineral premix.[72]

The logic of FDA's 1982 regulations for infant formula is apparent. Furthermore, these provisions were upheld by the U.S. Court of Appeals for the District of Columbia on December 31, 1985. FDA was supported in its interpretation of the act that "periodic testing" did not mandate batch-by-batch testing for each essential nutrient.[73] Congress, however, thought otherwise, and in 1986 it overrode FDA by passing an amendment to the act. Under this revision, testing is required to include the determination of vitamins A, B_1, C, and E at the final product stage for each batch. In addition, each premix must be checked for the relied-upon nutrients contained therein, and manufacturing controls must be established to test for all nutrients specified in the infant formula.[74]

Labeling and Promotion of Infant Formula

Concurrently with developing quality control regulations, FDA has considered revisions in the labeling of infant formulas. General concern has been expressed about the need to convey such information as directions for use to consumers, particularly those only speaking Spanish. The use of bilingual labels or the inclusion of a pictogram and symbols have been suggested. Also, proposals have been made to require instructions on the label for storage both before and after opening. The Infant Formula Council (IFC) representing manufacturers has offered advice to FDA on these issues. Out of this dialogue, FDA proposed a labeling revision on July 12, 1983. It provides for the declaration of all nutrients specified by the Infant Formula Act, expiration dating, and direction for use, including a pictogram.[75]

The criticism of infant formula has not been limited to the United States. For seven years an international boycott has been waged against Nestle, protesting its aggressive promotion of infant formula in less developed countries. Nestle, the world's principal supplier of infant formula outside the U.S. has been accused of disregarding the proper use of these products. Widespread illiteracy and poor sanitary conditions in these underdeveloped regions can lead to the misuse of formula causing serious illness. Backtracking from its original position, the company has agreed to comply with the voluntary guidelines of the World Health Organization. It has also reached an accord with protesters to:

1. curtail its promotional supplies of free infant formula to hospitals or health care centers in developing countries,
2. stop providing material favors to doctors in exchange for promoting the formula,
3. place warning labels on the packages, and
4. include warnings of the hazards of formula feeding in its promotional literature.[76]

PASTEURIZED MILK ORDINANCE

The Standard Milk Ordinance was introduced in 1924 by the U.S. Public Health Service to assist states and municipalities in managing a safe milk supply. This model regulation, now known as the Grade "A" Pasteurized Milk Ordinance (PMO), is available for adoption by the more than 15,000 state, county, and local health jurisdictions. The ordinance was developed and is currently updated with the assistance of milk sanitation and regulatory officials at every level of federal, state, and local government. These individuals are members of the National Conference of Interstate Milk Shipments (NCIMS) which has a Memorandum of Understanding with FDA prescribing their joint responsibilities in protecting the nation's milk supply. Milk products processed under the PMO are accepted as Grade A and can be so labeled. This grade is not to be confused with USDA grades. (See Chapter Three.)[77]

Pasteurization Conditions

The PMO specifies acceptable conditions for the manufacture of milk products. Of foremost concern is the temperature-time relationships for pasteurization, including the new process for UHT (ultra-high temperature) or aseptically packed milk. FDA requires that all interstate shipments of Grade A milk be pasteurized, but except for UHT milk these products do not have to be labeled as such. (Many state regulations, however, do require that "pasteurized" be printed on the label.) Homogenization is optional under federal regulation, but it has proven to be popular and is almost universal.[78] The PMO also gives rules governing, among other provisions, the inspection of milk haulers, tests for antibiotics, tamper-proof caps and closures, and sanitation practices. Former product definitions given in the PMO have been replaced by references to FDA standards of identity.[79] Approved suppliers of milk and dairy products are listed in the quarterly publication, *Sanitation Compliance and Enforcement Ratings of Interstate Milk Shippers* (IMS List).[80]

Enforcement of PMO

There was a time when FDA became disenchanted with the PMO and its enforcement. In 1975 FDA published a proposal that would have abandoned the PMO and put the milk industry under strict federal control. In the face of overwhelming opposition, FDA backed off from this position and went along with major PMO revisions that were finalized in 1979. This about-face, however, did not resolve all disputes. Smarting from growing imports of casein, the dairy industry has sought curbs on the use of this ingredient.

In 1981 the National Milk Producers Federation introduced a resolution at the NCIMS to "maintain the integrity of the Grade A designation." The proposed amendment to the PMO provided that, "Milk-derived ingredients which do not originate from an approved Grade A milk source shall not be permitted in milk and milk products . . . which are labeled Grade A." The motion was passed in spite of strong objections raised by the FDA representative who questioned the need for the change from a public health standpoint.[81] Six months later when the NCIMS executive board reaffirmed this action, an FDA spokesman equivocated, "In a case where the agency disagrees with a conference revision, it will either omit the change or include it and let the states decide whether they want to adopt and enforce it."[82] When FDA reached a final decision in May, 1982, it went along with the NCIMS resolution on prohibiting non-Grade A ingredients, including imported casein.[83]

INTERSTATE SHELLFISH SANITATION CONFERENCE

The head of quality assurance for Booth Fisheries has observed, "In the frozen fish and seafood industry, preservation of freshness is a struggle against time." He further stated that fishery products must be handled and processed as rapidly as possible, from the time they are taken out of the water until they are packed and preserved. Three factors are critical in retarding spoilage: time, temperature, and cleanliness. A maximum temperature of 40 °F (4 °C) is recommended during handling until freezing or preservation. Sanitation must begin with the fishing boats and receive major emphasis throughout processing. Compared with other food, seafood is extremely sensitive to abuse.[84]

National Marine Fisheries Service

A voluntary government inspection and grading service has been a great benefit to processors of fish and shellfish products. The Agricultural Marketing Act of 1946 delegated to USDA the authority to inspect and certify agricultural commodities. Seafood products were included in this authorization, but USDA never implemented the program for these products. The Fish and Wildlife Act of 1956 transferred all functions related to commercial fisheries to the Fish and Wildlife Service of the U.S. Department of Interior. Shortly thereafter a formal program was begun. Then in 1970 these activities were placed in a newly formed group, the National Marine Fisheries Service (NMFS) of the U.S. Department of Commerce.

When the seafood inspection program was initiated by the Fish and Wildlife Service, a "continuous inspection" policy was adopted. This approach later proved to be too inflexible and inefficient for many processors. Inspection fees, as mandated by legislation, were assessed on the basis of costs incurred regardless of the specific needs of manufacturers. Therefore, when NMFS took over the program, it instituted a major change. Henceforth processors were encouraged to assume full responsibility for quality control, thus freeing NMFS to devote its attention to matters of compliance. Guidelines were issued to help processors prepare formalized quality assurance plans for approval by NMFS.[85]

Participants in the inspection program administered by NMFS are allowed to identify their products by a "Packed Under Federal Inspection" (PUFI) mark. Inasmuch as FDA has jurisdiction over seafood products, NMFS relies on the Current Good Manufacturing Practice regulation to determine proper procedures. Canned fishery products must be processed in accordance with the GMP's for low-acid canned foods. NMFS will accept processing equipment that has been approved by USDA and additives included in USDA's "List of Chemical Compounds."[86]

Notwithstanding the apparent success of the voluntary inspection program under NMFS, Representative Byron L. Dorgan expressed concern over the "crazy quilt" of seafood regulation. In 1984 he indicated that he would introduce legislation in Congress to require mandatory inspection of all fish and seafood by USDA. He stated that this action was necessary to provide adequate protection to consumers. Even though the seafood industry is far more diverse and fragmented than the meat and poultry industries, Representative Dorgan expressed the belief that it "is time for all competing animal proteins to be inspected by the USDA under similar inspection procedures."[87]

National Shellfish Sanitation Program

Over the years shellfish have received special attention by regulators. Mollusks, including clams, mussels and oysters, are particularly susceptible to contamination in polluted waters. These organisms feed by filtering large volumes of water, and in the process they will consume any refuse that may be present, including human pathogens. The danger of contamination has grown as population centers have discharged increasing quantities of sewage into neighboring harbors and bays.

In 1925 a severe outbreak of typhoid fever occurred in the United States from eating bad oysters. This episode led to the establishment of the National Shellfish Sanitation Program (NSSP) to combat future occurrences. Under this program, polluted waters are monitored by testing for such indicator bacteria as *Escherichia coli* and fecal coliforms in order to provide warnings of the presence of typhoid, hepatitis, or other disease organisms. Clams, mussels, and oysters may be harvested only from approved growing waters, and they must meet microbiological standards at the wholesale level. Scallops and crustacean (crab, lobster, shrimp, crayfish) are not as prone to contamination and therefore less rigidly regulated.[88]

The enforcement of NSSP in recent years has been criticized as being loose and ineffectual. In 1973 the General Accounting Office issued a report finding fault with the operations of the program. FDA blamed the problem on a lack of teeth in the program, and in 1975 the agency proposed replacing NSSP with a new regulation that would have provided tightened control over the shell fish industry from shuckers to packers. The reaction to this proposal was strongly critical, and Congress stepped into the fray by voting a moratorium. Blocked from proceeding with its plans, FDA decided on a new course to delegate increasing responsibility to the states. This approach was more in tune with the way the U.S. Public Health Service had previously handled the shellfish program.[89]

The Interstate Shellfish Sanitation Conference (ISSC) was chartered in 1982 as a tripartite of state control officials, FDA representatives, and members of the shellfish industry. Later joined by NMFS, ISSC is patterned after the National Conference on Interstate Milk Shipments. In a Memorandum of Understanding with FDA, ISSC has agreed to adopt the NSSP Manual of Operations, inform FDA of infringements, and provide assistance in securing corrective measures. The NSSP guidelines will be revised and updated to serve as a model code for the states. FDA has agreed to "recognize the ISSC as the primary voluntary national organization of state shellfish regulatory officials that will provide guidance and counsel on matters for the sanitary control of shellfish."[90] A monthly publication, *Interstate Certified Shellfish Shippers List*, gives the names of approved suppliers of mollusk shellfish in the United States and certain foreign countries including Canada and Mexico.[91]

REFERENCES

1. Elizabeth M. Whelan, Fredrick J. Stare, *Panic in the Pantry*, Atheneum, New York 1976, p. 202.
2. Joseph P. Hile,"What Is the Future of GMP's?" *Food Product Development*, September, 1977, p. 100.
3. H. W. Walker, "Good Manufacturing Practices — Review and Discussion," *Food Technology*, May, 1971, pp. 64-66.
4. *Federal Register*, May 15, 1984, p. 20484.
5. "Frozen Raw Breaded Shrimp GMP," *Food Chemical News*, November 24, 1986, p. 2.
6. "FDA Issues Revised Food Good Manufacturing Practices As Final Rule," *Food Chemical News*, June 23, 1986, pp. 30-33.
7. "Who Regulates Water? FDA or EPA?" *Food Engineering*, November, 1978, p. 28.
8. "Expand GMP's to Cover Cacao, Confectionery, Shellfish," *Food Processing*, August, 1975, pp. 11, 12.
9. *Federal Register*, June 8, 1979, pp. 33238-33248.
10. "New GMP Policy Will Stress 'What,' Not 'How'," *Food Engineering*, April, 1981, pp. 15, 16.
11. "FDA Issues Revised Food Good Manufacturing Practices As Final Rule," *Food Chemical News*, June 23, 1986, pp. 30-33.
12. *Inspection Operations Manual*, TN 77-25, Food and Drug Administration, November 15, 1977.
13. Harold Hopkins, "A Greater Margin for Food Safety," *FDA Consumer*, September, 1974.
14. F. C. Majorack, "FDA's Quality Assurance Programs: Tools for Compliance," *Food Technology*, October, 1971, pp. 38-42.
15. "Retail Confectioners to Join FDA Quality Assurance Assistance Program," *Food Chemical News*, April 7, 1980, p. 34.
16. "Bakers Developing Quality Assurance Guidelines under FDA's IQAAP," *Food Chemical News*, July 7, 1980, p. 25.
17. "Voluntary Standards Set Tone for Sanitary Design, Construction of Equipment," *Baking Industry*, May, 1982, p. 90.
18. "FDA Administrative Practices and Procedures Are Published," *Food Processing*, July, 1975, p. 14.
19. "Angelotti Outlines Expanded Voluntary Quality Control with Label Legends," *Food Chemical News*, June 12, 1978, pp. 56-59.
20. Kevin Thompson, "Discretionary Inspection Legislation Is Passed by Congress after Last-Ditch Industry Efforts," *Meat Industry*, November, 1986, p. 8.

21. "Congress Approves Discretionary Meat Inspection Bill," *Food Chemical News*, October 27, 1986, pp. 34-35.
22. "Maximum Frequency of Reduced USDA Inspection to Be Twice Weekly," *Food Chemical News*, January 5, 1987, pp. 43-44.
23. "End of Mandatory USDA Prior Label Approval Predicted by Hibbert," *Food Chemical News*, March 9, 1987, p. 62.
24. "USDA Will Take Over All DOD Inspection Activities by 1979," *Food Chemical News*, April 10, 1978, p. 13.
25. "USDA to Take Over New Hampshire Meat and Poultry Inspection," *Food Chemical News*, July 10, 1978, p. 49.
26. "Legislation Suggested for State Plants Switching to USDA Inspection," *Food Chemical News*, December 10, 1979, p. 45.
27. "USDA General Counsel Says Quality Control Data Could Be Withheld under FOIA," *Food Chemical News*, November 5, 1979, p. 29.
28. Judy Rice, "Voluntary QC System for Meat/Poultry Plants," *Food Processing*, December, 1980, pp. 12, 14.
29. USDA's Final QC Program Permits Uniform Logo on Packages," *Food Product Development*, October, 1980, pp. 10, 12.
30. Upton Sinclair, *The Jungle*, The Heritage Press, New York, 1965, p. 60.
31. Neal Karlen with Jeff B. Copeland, "A 'Mystery Meat' Scandal," *Newsweek*, September 24, 1984, p. 31.
32. Wm. S. Stinson, "Tracing the Development of Sanitary Food Processing Equipment," *Food Processing*, October, 1978, pp. 214-223.
33. United States Department of Agriculture, *Accepted Meat and Poultry Equipment*, Bulletin MPI-2, U.S. Government Printing Office, Washington, D.C.
34. Jan Sneesby, "Flexible Dynamics V. Rigid Codes: the Case for Voluntary Standards," *Baking Industry*, March, 1981, pp. 70-73.
35. Stephanie C. Crocco, "Improved Sanitation, Reduced Costs from Automatic CIP," *Food Engineering*, July, 1977, pp. 58-60.
36. Ronald Jowitt, "Guidelines for Effective Sanitation," *Food Engineering*, January, 1981, pp. 66-69.
37. Steven Lentsch, "Selecting the Right Sanitizer . . . Are You Fooling Yourself?" *Food Engineering*, July, 1978, pp. 72-74.
38. Robert J. Swientek, "Cleaning/Sanitation Quality Control Seminar," *Food Processing*, June, 1980, pp. 122-124.
39. *Sanitizing Solutions: Chemistry Guidelines for Food Additive Petitions*, Division of Food Chemistry & Technology, Food & Drug Administration, Washington, D.C. 20204, July, 1986.
40. Victor H. Durler, "Sanitizing Without Chlorine: Quats, Iodophors and Acid-Anionics," *Food Engineering*, January, 1978, p. 128.
41. Toni Antonetti, "Baking Industry Guide: Sanitizing Chemicals," *Baking Industry*, March, 1979, pp. 30, 31.

42. Klaus Buehring, "Sanitation Integrated with Production Functions," *Baking Industry*, August, 1978, pp. 8-10.
43. C. T. Cowan, "Corrosion of Stainless Steel . . . How to Prevent It," *Food Engineering*, July, 1977, pp. 64-67.
44. John Forwalter, "1980 Selection Guide Cleaning and Sanitizing Compounds," *Food Processing*, February, 1980, pp. 40-43.
45. Floyd Bodyfelt, "97% Clean Is Still 3% Dirty," *Dairy Record*, July, 1979, p. 80.
46. "Tips on Cleaning-in-Place," *Food Engineering*, July, 1980, pp. 94-98.
47. Barry Landa, "Water Conservation: How Will It Affect Sanitation?" *Food Engineering*, July, 1977, p. 57.
48. C. F. Bryson, Jr., "Needed: Shared Sanitation Technology," *Food Engineering*, July, 1977, p. 61.
49. "Review of Processors' 'Public Enemies' Determines Frequently Encountered Pests," *Food Engineering*, December, 1977, p. 147.
50. Robert Bannar, "Sanitation: More Than a Daily Cleanup," *Food Engineering*, July, 1978, pp. 63, 64.
51. Allen M. Katsuyama, "Practical Ideas for Improving Plant Sanitation," *Food Engineering*, June, 1979, pp. 107-109.
52. Lee Barksdale, "How to Control Pests," *Food Engineering*, July, 1980, pp. 106-108.
53. Kenneth V. Nyberg, "Uncover Weaknesses in Your Sanitation Program," *Food Engineering*, July, 1977, pp. 62, 63.
54. John Forwalter, "EPA Pesticide Classification Continues; USDA Issues Meat & Poultry Plant Chemicals List," *Food Processing*, February, 1980, pp. 52, 53.
55. "Bad Humor," *Newsweek*, August 18, 1975, p. 65.
56. "Ice Cream Gate." *Time*, August 18, 1975, p. 67.
57. Robert L. Lyons, "Required Personnel Training in Plant Quality Control," *Food Technology*, April, 1984, pp. 105-110.
58. Food and Drug Administration, "Food Industry Education Programs (FY79)," *Compliance Program Guidance Manual*, 7303.833, Washington, D.C.
59. "Reorganizes Bureau of Foods," *Food Technology*, May, 1984, p. 56.
60. Food and Drug Administration, *FDA Manuals for the Food and Cosmetic Industry*, DHEW Publication No. (FDA) 79-1060, Washington, D.C.
61. Food and Drug Administration, *FDA's Catalog of Information Materials for the Food and Cosmetic Industries*, DHEW Publication No. (FDA) 80-1067, Washington, D.C.
62. Food and Drug Administration, *So You Work in a Food Plant!* HHS Publication No. (FDA) 72-2032, Washington, D.C.
63. Food and Drug Administration, *Follow the Signs to Safe Food*, DHHS Publication No. (FDA) 80-2133, Washington, D.C.

64. Floyd W. Bodyfelt, "Personnel As a Source of Microbial Contamination," *Dairy Record*, October, 1981, pp. 250, 252.
65. Rafael R. Pedraja, "People: The Key to Effective Sanitation," *Food Engineering*, July, 1978, pp. 76-78.
66. Eliot Marshall, "The Enduring Problem of Pesticide Misuse," *Technology Review*, February/March, 1984, pp. 10, 11.
67. J. W. James, "Good Laboratory Practice," *Chemtech*, March, 1982, pp. 162-165.
68. Annabel Hecht, "New Standards for Test Laboratories," *FDA Consumer*, March, 1977, pp. 22-25.
69. Harold Hopkins, "Next to Mother's Milk, There's Infant Formula," *FDA Consumer*, July-August, 1980, pp. 11-13.
70. "Infant Formula Mishap Sparks Legislation," *Processed Prepared Food*, March, 1981, pp. 28, 31.
71. "Flexibility Foremost Feature of Final Infant Formula Quality Control Rule," *Food Chemical News*, April 26, 1982, pp. 26-32.
72. "FDA Proposes Quality Control Rules for Infant Formulas," *Food Chemical News*, January 12, 1981, pp. 7-14.
73. "Appeals Court Upholds FDA Infant Formula Quality Control Rules," *Food Chemical News*, January 13, 1986, pp. 16-20.
74. "Infant Formula Exemption Continuation Requests Needed," *Food Chemical News*, November 17, 1986, pp. 41-43.
75. "Proposes Infant Formula Labeling Requirements," *Food Technology*, September, 1983, p. 20.
76. "Nestle Boycott Being Suspended," *The New York Times*, January 27, 1984, pp. A1, A4.
77. Donald Kennedy, "Statement before the Subcommittee on Dairy and Poultry, Committee on Agriculture," Washington, D.C., May 23, 1977.
78. Chris Lecos, "Milk Cows Produce It; Man Improves It," *FDA Consumer*, June, 1982, pp. 16-20.
79. "Copies of Revised PMO Are Made Available for Comment by FDA," *Food Chemical News*, February 7, 1977, pp. 37, 38.
80. *Sanitation Compliance and Enforcement Ratings of Interstate Milk Shippers*, Public Health Service, Food and Drug Administration, HFF-346, Washington, D.C. 20204, April 1, 1987.
81. "Settle Casein Question with PMO Amendment?" *Dairy Record*, July, 1981, p. 37.
82. "Board Forbids Non-Grade A Ingredients in Milk Products, Despite FDA's Warning," *Dairy Record*, January, 1982, p. 25.
83. "Final PMO Amendments Leave UHT Labeling to Processor," *Dairy Record*, May, 1982, p. 26.

84. Rafael R. Pedraja, "Dynamic Sanitation in the Fish and Seafood Industry," *Food Technology*, October, 1973, pp. 42, 44, 72.
85. Irving D. Sackett, Jr., "Quality Inspection Activities of the National Marine Fisheries Service," *Food Technology*, June, 1982, pp. 91, 92.
86. "National Marine Fisheries Service," *Food Processing*, July, 1976.
87. "Bill to Make Fish Inspection Mandatory under USDA Pledged by Dorgan," *Food Chemical News*, February 20, 1984, pp. 46-48.
88. Jack R. Matches and Carlos Abeyta, "Indicator Organisms in Fish and Shellfish," *Food Technology*, June, 1983, pp. 114-117.
89. "FDA May Abandon Shellfish Sanitation Regulations," *Food Chemical News*, April 17, 1978, pp. 3, 4.
90. "Shellfish Sanitation Conference Formalized in Agreement with FDA," *Food Chemical News*, April 2, 1984, pp. 27, 28.
91. *Interstate Certified Shellfish Shippers List*, Public Health Service, Food and Drug Administration, HFF-344, Washington, D.C. 20204, April 1, 1987.

CHAPTER SIX
PLANT INSPECTION

Enforcement of the food laws depends on the authority given to government agencies to inspect processing facilities. The U.S. Department of Agriculture (USDA) is responsible for inspection over meat, poultry, and egg products, but the vast majority of food plants come under the jurisdiction of the Food and Drug Administration (FDA). Plant inspections are essential to consumer protection. These inspections, however, are of only limited value if both regulator and manufacturer confront each other as adversaries. Instead, what is required is a spirit of cooperation so that problems can be aired frankly and resolved to everyone's satisfaction. In order that an inspection be conducted in a cordial and correct manner, both parties, the FDA inspector and the company representative, need to be cognizant of each other's rights and obligations under the Federal Food, Drug, and Cosmetic Act (FD&C Act).

FDA INSPECTIONAL AUTHORITY

FDA is vested with plenipotentiary authority to enter unannounced any factory, warehouse, or establishment where food is processed, packed, or held for introduction into interstate commerce or after such introduction. This inspectional authority extends to any vehicle being used to transport or hold such food. FDA wields these powers notwithstanding the Fourth Amendment to the U.S. Constitution which reads in part, "The right of the people to be secure in their persons, houses, papers, and effects, against unreasonable searches and seizures, shall not be violated. . . ." In the interest of food safety, Congress has seen fit to make special provisions for factory inspection. Section 704 of the FD&C Act stipulates that FDA representatives

> . . . upon presenting appropriate credentials and a written notice . . . are authorized . . . to inspect, at reasonable times and within reasonable limits and in a reasonable manner, such factory, warehouse, establishment, or vehicle and all pertinent equipment, finished and unfinished materials, containers, and labeling therein.

153

Warrantless Inspections

The authority of FDA to make inspections without receiving a court issued warrant has been debated and contested over the years. The FD&C Act passed in 1938 originally authorized factory inspections, but this provision was curtailed by a U.S. Supreme Court decision in 1952. Thereupon in 1953 Congress adopted the language of Section 704 as an amendment to the Act, giving FDA explicit authority to conduct warrantless inspections without the factory owner's permission. Furthermore, under Sections 301(f) and 303(a) of the FD&C Act "the refusal to permit entry or inspection as authorized by Section 704" is a prohibited act and therefore punishable by fines and/or imprisonment. The federal courts have repeatedly upheld the constitutionality of the 1953 amendment on the basis that the food industry is "pervasively regulated" and thus forfeits claims of privacy.[1]

Regulations have not been promulgated to prescribe when inspections are to be instituted or how they are to be conducted. To the dismay of some individuals FDA has wide latitude in its approach to inspections. Accusations have been made that the inspectors' manual endorses "gamesmanship."[2] For example, the inspector is told, "ask to review the firm's complaint files," even though these records are off-bounds to FDA. In a rejoinder FDA has testified that it does not normally use surreptitious surveillance or investigative techniques. Recording devices are specifically prohibited.[3] The agency, however, considers random, unannounced inspections an essential part of its compliance scheme. It also defends its stance that its inspectors are free to ask for more information than they are entitled to receive under the statute. Moreover, an inspector is not obligated to give a *Miranda* warning to a manufacturer reminding him of his right to refuse queries on grounds of self-incrimination.[4]

Types of Inspections

However capricious inspections may appear to food manufacturers, there generally is a logical and orderly pattern to them. Frequently inspections are triggered by customer complaints or reports of violations. Water G. Campbell, Chief of FDA from 1927 until his retirement in 1944, instituted the approach that the detection of a violation should be, as nearly as possible, at the source of the violation.[5] Concurrent with its compliance activities, FDA will conduct routine inspections geared to mutual enlightenment and assistance. In-depth HACCP inspections of facilities are conducted in order to determine the potential hazards and to ascertain the critical control points needed to minimize these hazards. In these types of inspections, FDA frequently advises the manufacturer of its intentions in advance and arranges a preinspection conference for the purpose of familiarizing the inspectors with the process.

Another inspectional approach used by FDA is the System Inspection which has been detailed by Donald C. Healton, Executive Director of Regional Operations. (EDRO) This procedure affords a minute look at one or more systems which have a direct bearing on the quality of the food products. Many such systems exist and are candidates for thorough reviews. Examples are master formula/batch records controls, laboratory/testing procedures, packaging and labeling controls, plumbing facilities, and ventilating systems.[6]

FDA administers two important inspection programs in connection with other agencies. At the urging of the Office of Management and Budget (OMB) FDA implemented in 1974 the Government-Wide Quality Assurance Program for the procurement of drugs by federal departments. Prior to awarding a contract, the evaluation and rating of suppliers is critical to this program. In order to rate a manufacturer, FDA inspects the firm's facilities and then prepares a profile report for quick reference. Because of the success of the program, consideration has been given to extending it to food products purchased by the government.[7]

The second surveillance activity of significance is FDA's State Contract Program. In fiscal year 1977 there were 33 states participating in this joint program to provide inspection of food establishments. These facilities are generally involved in both interstate and intrastate commerce. Not only does the program permit a reduction in duplicate inspectional efforts, but FDA concedes that in many instances state agencies can move more swiftly and effectively than the federal government. Armed, in many cases, with powers to embargo adulterated food, halt sales of questionable products, withdraw licenses or permits, and impose fees, state health departments have played an important part in raising the general level of consumer protection.[8]

Inspectional Routine

A typical inspection by FDA begins when an agency representative knocks on the manufacturer's door. As specified by law, the inspection must be conducted at "reasonable times," which has been interpreted as during normal business hours or whenever a facility is in production, not excepting Saturdays, Sundays, holidays, and nights.[9] The inspector must submit his credentials as well as a Notice of Inspection or Form FD482 (Figure 6.1). If non-FDA persons are accompanying the inspector, the manufacturer has a right to deny access to those persons without valid credentials. At the outset the plant should feel free to inquire about the nature of the inspection, the products or processes to be covered, and the reasons for the visit, e.g., a customer complaint. The plant should also advise the inspector of any and all safety precautions and plant rules.

A person of authority and with knowledge of the operation should accompany the inspector at all times throughout the tour of the plant. Generally this individual is the plant manager. Other staff personnel may also be present, including the quality control manager and the production superintendent. In a small business the owner or president most likely will want to participate in the discussions. The top ranking officer representing the firm should be prepared to decide what questions to answer, what documents and records to show, and what infor-

DEPARTMENT OF HEALTH AND HUMAN SERVICES
PUBLIC HEALTH SERVICE
FOOD AND DRUG ADMINISTRATION

1. DISTRICT ADDRESS & PHONE NO.
Rm 563 Federal Office Bldg.
30 U.N. Plaza
San Francisco, CA 94102
(415) 556-2062

2. NAME AND TITLE OF INDIVIDUAL
Robert K. Thompson, Plant Manager

3. DATE
5-15-85

TO

4. FIRM NAME
Garden City Nut Shellers

5. HOUR
8:30 a.m.

6. NUMBER AND STREET
2704 Sellers Ave.

p.m.

7. CITY AND STATE & ZIP CODE
San Jose, CA 95131

8. PHONE # & AREA CODE
(408) 123-4567

Notice of Inspection is hereby given pursuant to Section 704(a)(1) of the Federal Food, Drug, and Cosmetic Act [21 U.S.C. 374(a)][1] and/or Part F or G, Title III of the Public Health Service Act [42 U.S.C. 262-264][2]

9. SIGNATURE *(Food and Drug Administration Employee(s))*
Sidney H. Rogers

10. TYPE OR PRINT NAME AND TITLE *(FDA Employee(s))*
Sidney H. Rogers
Investigator

Applicable portions of Section 704 and other Sections of the Federal Food, Drug, and Cosmetic Act [21 U.S.C. 374] are quoted below:

[1]Sec. 704. (a)(1) For purposes of enforcement of this Chapter, officers or employees duly designated by the Secretary, upon presenting appropriate credentials and a written notice to the owner, operator, or agent in charge, are authorized (A) to enter, at reasonable times, any factory, warehouse, or establishment in which food, drugs, devices, or cosmetics are manufactured, processed, packed, or held, for introduction into interstate commerce or after such introduction, or to enter any vehicle being used to transport or hold such food, drugs, devices, or cosmetics in interstate commerce; and (B) to inspect, at reasonable times and within reasonable limits and in a reasonable manner, such factory, warehouse, establishment, or vehicle and all pertinent equipment, finished and unfinished materials, containers, and labeling therein. In the case of any factory, warehouse, establishment, or consulting laboratory in which prescription drugs or restricted devices are manufactured, processed, packed, or held, the inspection shall extend to all th... [including records, files, papers, processes, controls, bearing on whether prescription drugs or restrict... adulterated or misbranded within the meaning ... may not be manufactured, introduced in... sold, or offered for sale by reason of a... been or are being manufactured, ... held in any such place, or otherwise ... No inspection authorized by the pre... (3) shall extend to financial data, sale... pricing data, personnel data *(other th... technical and professional personnel p... this Act)*, and research data *(other than... antibiotic drugs and devices and, subject... under regulations lawfully issued pursuan... section 507(d) or (g), section 519, or 52... other drugs or devices which in the case of ... ject to reporting or inspection under lawful n... to section 505(k) of the title. A separate noti... such inspection, but a notice shall not be requi... during the period covered by the inspection. Each such inspection shall be commenced and completed with reasonable promptness.

Sec. 704(e) Every person required under section 519 or 520(g) to maintain records and every person who is in charge or custody of such records shall, upon request of an officer or employee designated by the Secretary, permit such officer or employee at all reasonable times to have access to and to copy and verify, such records.

Section 512 (1)(1) In the case of any new animal drug for which an approval of an application filed pursuant to subsection (b) is in effect, the applicant shall establish and maintain such records, and make such reports to the Secretary, of data relating to experience and other data or information, received or otherwise obtained by such applicant with respect to such drug, or with respect to animal feeds bearing or containing such drug, as the Secretary may by general regulation, or by order with respect to such application, prescribe on the basis of a finding that such records and reports are necessary in order to enable the Secretary to determine, or facilitate a determination, whether there is or may be ground for invoking subsection (e) or subsection (m)(4) of this section. Such regulation or order shall provide, where the Secretary deems it to be appropriate, for the examination, upon request, by the persons to whom such regulation or order is applicable, of similar information received or otherwise obtained by the Secretary.
(2) Every person required under this subsection to maintain records, and every person in charge or custody thereof, shall, upon request of an officer or employee designated by the Secretary, permit such officer or employee at all reasonable times to have access to and copy and verify such records.

[2]Applicable sections of Parts F and G of Title III Public Health Service Act [42 U.S.C. 262-264] are quoted below:

Part F - Licensing — Biological Products and Clinical Laboratories and******

Sec. 351(c) "Any officer, agent, or employee of the Department of Health & Human Services, authorized by the Secretary for the purpose, may during all reasonable hours enter and inspect any establishment for the propaga... or manufacture and preparation of any virus, serum, toxin ... vaccine, blood, blood component or derivative, all... or other product aforesaid for sale, barter, or exch... f Columbia, or to be sent, carried, or brought ...n into any other State or possession or into ...m any foreign country into any State or ...tion.

...nds for good cause that the methods, ...tronic product radiation safety in a ...se, or establishment in which electronic ...ured or held, may not be adequate or reliable, ...es duly designated by the Secretary, upon presenting ...edentials and a written notice to the owner, operator, or ... charge, are thereafter authorized (1) to enter, at reasonable ...s any area in such factory, warehouse, or establishment in which the manufacturer's tests *(for testing programs)* required by section 358 (h) are carried out, and (2) to inspect, at reasonable times and within reasonable limits and in a reasonable manner, the facilities and procedures within such area which are related to electronic product radiation safety. Each such inspection shall be commenced and completed with reasonable promptness. In addition to other grounds upon which good cause may be found for purposes of this subsection, good cause will be considered to exist in any case where the manufacturer has introduced into commerce any electronic product which does not comply with an applicable standard prescribed under this subpart and with respect to which no exemption under section 359(a)(2) or 359(e)."

(b) "Every manufacturer of electronic products shall establish and maintain such records *(including testing records)*, make such reports, and provide such information, as the Secretary may reasonably require to enable him to determine whether such manufacturer has acted or is acting in compliance with this subpart and standards prescribed pursuant to this subpart and, upon request of an officer or employee duly designated by the Secretary, permit such officer or employee to inspect appropriate books, papers, records, and documents relevant to determining whether such manufacturer has acted or is acting in compliance with standards prescribed pursuant to section 359(a)."

******

(f) "The Secretary may by regulation (1) require dealers and distributors of electronic products, to which there are applicable standards prescribed under this subpart and the retail prices of which is not less than $50, to furnish manufacturers of such products such information as may be necessary to identify and locate, for purposes of section 359, the first purchasers of such products for purposes other than resale, and (2) require manufacturers to preserve such information.

FORM FDA 482 (5/85)   PREVIOUS EDITION IS OBSOLETE   NOTICE OF INSPECTION

(watermark: SAMPLE USE ONLY NOT AN OFFICIAL ISSUANCE)

FIGURE 6.1. FORM FDA 482, NOTICE OF INSPECTION

mation to claim as privileged.[10] On occasion the company spokesman may ask the inspector's indulgence to defer a reply until senior management has reviewed a specific request.

Under the law, FDA is not granted the authority to inspect food records. This restriction is in stark contrast to the legal right given to search the records, files, papers, processes, controls, and facilities related to the manufacture of prescription drugs. In food plants, however, the agency may not inspect, without the owner's permission, any financial records, sales figures other than shipping data, pricing information, personnel files other than data relating to the qualifications of professionals, and research reports. In floor debate on the 1953 amendment to the FD&C Act, Congress clarified that access was not being given to such records as product formulas, raw material specifications, complaint files, or quality control records.

In spite of the above restrictions on what information FDA is entitled to see, there are several important exceptions. Section 704 of the FD&C Act provides for the inspection of any records bearing on whether infant formula complies with the requirements of the law. The regulation for Emergency Permit Control (21CFR108.25[g], 21CFR108.35[h]) gives permission to an inspector, upon written notice, to review and copy all processing records pertaining to acidified foods and to thermally processed, low-acid canned foods.[11] Finally, a manufacturer might as well volunteer reports of analysis on finished products which have been cleared for shipment since this information can be retrieved upon written request from shippers and carriers pursuant to Section 703 of the Act.

The Question of Photographs

FDA maintains that the FD&C Act gives it tacit permission to take photographs during the inspections of food establishments. In an often quoted case, *U.S. v. ACRI Wholesale Grocery*, the U.S. District Court for the Southern District of Iowa found in 1976 that "the photographing of warehouse conditions by FDA agents was not unreasonable."[12] Some legal opinion, however, differs with this conclusion. Independent counsel cautions that photographs used as evidence often take a violation out of context. For example, by focusing on two contaminated food items out of a lot of 1,100 units, a photograph can distort actual conditions. Therefore some lawyers have advised their clients to insist that FDA inspectors obtain warrants before photographs are taken.[13] In any event, manufacturers should request that FDA furnish to them duplicate copies of all photographs taken during an inspection.

Photographs, files, or other records received by FDA may contain proprietary information that should be protected. Indiscriminate assertions of trade secrets, however, are unjustified and are frowned upon by FDA. Nevertheless, legitimate claims of secrecy are protected under the federal confidentiality statute (18 U.S.C.1905) and the FD&C Act (Section 301[j]). Any information not designated as proprietary will be released to the public under the Freedom of Information Act (FOI). The responsibility rests with the manufacturer to inform FDA of the confidentiality of specific information. A recommended policy for companies is to physically segregate all material to which FDA may have access so that there can be no inadvertent disclosure.[14] Separate log books should be kept for laboratory data, one set open for inspection and the other to be held in confidentiality.

Samples of Materials

Explicit permission is given to FDA by statute to take samples of materials. Before the inspector leaves the establishment, a receipt describing the samples taken must be given to the manufacturer. Form FD484, Receipt For Samples, (Figure 6.2) has been prepared for this purpose. FDA will reimburse a company for the cost of the samples, the usual procedure being for the company to bill FDA's district office. A copy of FDA's analyses of a sample will promptly be furnished to the manufacturer. Whenever an FDA inspector takes a sample, the manufacturer is advised to take the following steps:

1. Ask the inspector the reason for taking the sample and the tests to be performed on it.
2. Put on hold the lot from which the sample was taken.
3. Seal any rail car that has been inspected by FDA and from which samples may have been taken until its status has been determined.
4. Immediately notify customers of any product shipped to them from the sampled lots.
5. Retain duplicate or split samples for future reference.
6. Obtain analyses on the sample from an independent testing laboratory.

Besides samples, an FDA inspector may request other materials. For example, he is entitled to copies of labels and promotional literature. He need not sign for the receipt of this printed matter. As a precaution, if an inspector takes any labels, they should be marked "specimen," and the manufacturer should retain duplicate ones for his own records. Information on product coding and a list of consignees of interstate shipments should be supplied upon request. A copy of the firm's recall procedure should also be made available.

Throughout an inspection, the manufacturer should take detailed notes on the remarks of the inspector and any criticism or advice he may offer. A common practice worth following is to "red tag" any physical facility considered to be a problem. If at all possible, and management is in accord, any noted violation should be immediately corrected while the inspector is present. Many careless or overlooked defects can be remedied on the spot.

Written Reports

At the conclusion of an inspection, the FDA official is required by law to leave with the manufacturer a written report setting forth any conditions or practices which, in the inspector's judgement, do not conform to the regulations. Such a report is presented on Form FD483, titled Inspectional Observations (Figure 6.3). If a Form 483 is not left by the inspector, this fact is evidence that he found nothing wrong.

It is imperative that the manufacturer take the contents of a 483 report seriously. He should review the report with the inspector before the latter departs. Any

DEPARTMENT OF HEALTH AND HUMAN SERVICES	1. DISTRICT ADDRESS		
PUBLIC HEALTH SERVICE	Food & Drug Administration		
FOOD AND DRUG ADMINISTRATION	585 Commercial St. Boston, MA 02)09		

2. NAME AND TITLE OF INDIVIDUAL		3. DATE	4. SAMPLE NUMBER
Mr. Richard R. Frost, General Mgr.		*10-5-84*	*85-123-543*

5. FIRM NAME	6. FIRM'S DEA NUMBER	7. FDA'S DEA NUMBER
Quality Wholesale Drug Co.	*AB 3632918*	*PD 0052112*

8. NUMBER AND STREET	9. CITY AND STATE (Include Zip Code)
3146 Front St.	*Boston, Ma. 02107*

10. SAMPLES COLLECTED (Describe fully. List lot, serial, model numbers and other positive identification)

The following samples were collected by the Food and Drug Administration and receipt is hereby acknowledged pursuant to Section 704(c) of the Federal Food, Drug, and Cosmetic Act ☐21 U.S.C. 374(c)☐ and or Part F, Sub Part 3, Section 356(b) of The Public Health Service Act ☐42 U.S.C. 263d☐ and/or 21 Code of Federal Regulations (CFR) 1307.02. Excerpts of these are quoted on the reverse of this form.

One box of 25-1cc ampules, Dilaudid HCl (hydromorphone) 2mg/cc, Lot # 0103 213 manufactured by Knoll Pharmaceutical company, Orange, NJ.

```
**********************
*************** ONLY
********* SAMPLE USE
****** NOT AN OFFICIAL ISSUANCE ******
*************
```

11. SAMPLES WERE	12. AMOUNT RECEIVED FOR SAMPLE	13. SIGNATURE (Person receiving payment for sample)
☒ PURCHASED ☐ BORROWED (To be returned)	*$5.00* ☒ CASH ☐ BILLED ☐ VOUCHER	*Dealer affidavit signed*

14. COLLECTOR'S NAME (Print or Type)	15. COLLECTOR'S TITLE (Print or Type)	16. COLLECTOR'S SIGNATURE
SIDNEY H. ROGERS	*INVESTIGATOR*	*Sidney H. Rogers*

FORM FDA 484 (4/74) PREVIOUS EDITION MAY BE USED.   RECEIPT FOR SAMPLES   PAGE 1 OF 1 PAGES

FIGURE 6.2. FORM FDA 484, RECEIPT FOR SAMPLES

misunderstanding should be clarified. If the manufacturer disagrees with any part of the report he might, after supplying additional information, get the inspector to change it or strike it from the record. Otherwise the manufacturer should request the inspector to note the points of disagreement. In addition, the FDA

DEPARTMENT OF HEALTH AND HUMAN SERVICES PUBLIC HEALTH SERVICE FOOD AND DRUG ADMINISTRATION	DISTRICT ADDRESS AND PHONE NUMBER Minneapolis District 240 Hennepin Ave. Minneapolis, MN 55401 (612) 787-3904	
NAME OF INDIVIDUAL TO WHOM REPORT ISSUED TO: William S. Gundstrom	PERIOD OF INSPECTION Jan. 5 - 7, 1985	C. F. NUMBER
TITLE OF INDIVIDUAL Vice President Production	TYPE ESTABLISHMENT INSPECTED Tablet Repacker	
FIRM NAME Topline Pharmaceuticals "T.L.P."	NAME OF FIRM, BRANCH OR UNIT INSPECTED "T.L.P." DIVISION 3	
STREET ADDRESS 2136 Elbe Place	STREET ADDRESS OF PREMISES INSPECTED 80 Elbe Court	
CITY AND STATE (Zip Code) Jackston, MN 55326	CITY AND STATE (Zip Code) North Jackston, MN 55327	

DURING AN INSPECTION OF YOUR FIRM (I) WAS OBSERVED:

List your observations in a logical and concise manner.

****
SAMPLE USE ONLY
NOT AN OFFICIAL ISSUANCE
****

SEE REVERSE OF THIS PAGE	EMPLOYEE(S) SIGNATURE *Sidney H. Rogers*	EMPLOYEE(S) NAME AND TITLE (Print or Type) Sidney H. Rogers Investigator	DATE ISSUED 1-7-85

FORM FDA 483 (5/85)   PREVIOUS EDITION MAY BE USED.   INSPECTIONAL OBSERVATIONS PAGE 1 OF 1 PAGES

FIGURE 6.3. FORM FDA 483, INSPECTIONAL OBSERVATIONS

representative should be asked to include in his report any corrective actions taken in his presence.

A copy of the 483 report which is given to the plant is now routinely sent to the attention of top management, i.e., the chief executive officer of the corporation. Although the FD&C Act restricts the inspector's report to observations involving filth, decomposition, and insanitation, in reality the use of the 483 form has been expanded to cover a multitude of conditions. This broader coverage was initially well received by industry inasmuch as these reports were not circulated. Since 1974, however, FDA has made all 483 reports available upon request under FOI.

While not required to do so by the regulations, the company whose premises have been inspected is encouraged to file at its earliest convenience a formal response to any 483 report. This memorandum should be addressed to FDA's district office, and it should include a summary of corrective measures already taken and a plan of action to fix all remaining problems thought to be valid. If there are questions these should be raised, and where areas of disagreement exist, they should be noted and substantiated. Such a report should be prepared with the same care as would be required for any legal document. The policy of FDA is to acknowledge all correspondence relating to 483 reports. The agency believes that corporate responses are conducive to effective regulation and favors this approach.[15]

An Establishment Inspection Report (EIR) is prepared based on the contents of the 483 report, any rebuttal by the manufacturer, other available evidence, and background information on the company and its business. This report is put together for internal use by FDA. The purpose of the EIR is to present a frank and critical assessment of the plant's compliance status. In addition to the inspector's evaluation, the report will include comments from the district reviewing officer. Unlike the 483 report, the EIR is not immediately available to the public. At such time that FDA determines that no regulatory action will be taken, the non-confidential sections of the EIR are released.

The inspectional authority accorded FDA does not extend to independent testing laboratories. Only those laboratory facilities that are an integral part of the manufacturing plant are subject to inspection under the FD&C Act. This restriction has posed a dilemma to FDA concerning the enforcement of the Good Laboratory Practice regulations. After being confronted with several refusals of entry, FDA has been forced to take the position that studies submitted by laboratories which bar inspectors will not be accepted in support of registration petitions.[16]

SEIZURE, INJUNCTION, AND PROSECUTION

The comment has been made that "Inspectional Observations" is a misnomer for the 483 report. It really should be called "Inspectional Violations." Inevitably a Form 483 report is the basis for follow-up action by FDA. The course of action taken by the agency will depend on the seriousness of the violation and the attitude of the firm with regard to the warning. Other factors will also come into play, such as the cost-effectiveness of any legal proceedings, not to mention FDA's desired public image ("reasonableness rather than . . . vindictiveness").[17]

The FC&C Act gives FDA considerable flexibility in enforcement decisions. Section 306 allows that:

> Nothing in this Act shall be construed as requiring the Secretary to report for prosecution, or for the institution of libel or injunction proceedings, minor violations of this Act whenever he believes that the public interest will be adequately served by a suitable written notice or warning.

"Notice of Adverse Findings" and "Regulatory Letter"

With this provision as justification, FDA has formalized two kinds of written response to disclosures of violations. First described in a 1978 proposed regulation, a Notice of Adverse Findings was designed to replace the old and somewhat ambiguous "reports of inspectional findings," otherwise known as "information letters." The second type of warning, which is sent in more serious cases, was called a Regulatory Letter.[18] After receiving heated criticism on details of its proposal, FDA eventually decided to withdraw the regulation and amend instead its Regulatory Procedures Manual (RPM) by including guidelines for the issuance of these letters.

The RPM guidelines provide that a Notice of Adverse Findings will be issued when:

(a) Agency policy is clear with respect to the condition, practice, or violation; and

(b) the agency has information that a firm, product, and/or individual is in violation of the law or regulations or there is information that an existing condition or practice may lead to a violation if left uncorrected; and

(c) the agency has concluded that the nature of the violation does not require immediate action directed against the firm and/or individual(s).

A Regulatory Letter will be issued under the following circumstances:

(a) Agency policy is clear with respect to the condition, practice, or violation;

(b) it has evidence that a firm, product and/or individual is in violation of the law or regulations and the agency has concluded that the violation(s) warrants the initiation of administrative or legal sanction(s). . .; and

(c) FDA is committed to initiate such administrative or legal action immediately if correction is not promptly achieved.

FDA's Discretionary Authority

On several occasions FDA has reiterated its position that it is not under obligation to issue written notices or warnings prior to taking legal action. A decision not to issue a letter but rather to resort to other measures will be determined by the following factors:

    (1) The violation reflects a recent history of repeated or continuous conduct of a similar or substantially similar nature during which time the firm and/or individual(s) have been notified of a similar or substantially similar violation: or

    (2) the violation is intentional or flagrant; or

    (3) the violation represents a reasonable possibility of injury or death.[19]

Regulatory Letters will specify which sections of the law or regulations have been broken, provide a brief description of the facts supporting the assertion, and request corrective action. Notices of Adverse Findings are to describe succinctly the condition or practice which is in violation, or may lead to a violation, and request corrective action. Copies of either type of letter are to be sent to the individual most closely connected with the infraction, this person's immediate superior, and the highest known official in the business. Whereas a Regulatory letter, after editing by FDA, will be released to the public, a Notice will be held confidential. Prompt responses are expected from either of these two kinds of communication. As a rule, FDA requests that a recipient of a Notice answer it within 30 days after its receipt, and in the case of a Regulatory Letter within 10 days.

Acceptable responses to Regulatory Letters should include:

    (1) . . . Each step that has been taken or is being taken to correct the violation(s);

    (2) . . . an explanation of each step being taken to prevent the recurrence of similar violations;

    (3) . . . if corrective action is not completed within the time period stipulated . . . [a statement of] the reason for the delay and the time within which the correction(s) will be completed.[20]

Similar responses should be submitted for Notices of Adverse Findings. In an effort to highlight "voluntary corrective action" by industry, FDA instituted in 1981 a system for collecting statistics on the number of actions. To qualify as a voluntary corrective action, the following criteria must be met.

- Detection of a problem(s)
- Correction of a problem(s)
- Verification of correction of a problem(s)[21]

Voluntary compliance has proven to be the rule rather than the exception, notwithstanding press releases giving the opposite impression. In the vast majority of cases, violations are immediately corrected in response to 483 reports, thus obviating the need to issue Regulatory Letters or Notices of Adverse Findings. The resolution of problems through voluntary means is much preferred by FDA which is more interested in securing compliance than punishing culprits. Where a proper response is not forthcoming from a manufacturer, however, FDA will resort to administrative and/or judicial sanctions, including seizure, injunction, license suspension or revocation, civil fines, and criminal prosecution.

Criminal Prosecution

The Pure Food and Drugs Act of 1906 provided that offending products could be seized and condemned persons fined or jailed. The regulatory authorities soon found that seizure and criminal prosecution are both time-consuming procedures. They necessitate going to court, and while a case is pending the accused is free to continue his malfeasance and dispose of violative product. Furthermore, seizure often has little deterrence because the value of the goods confiscated is not sufficiently great to have an economic impact. Strapped with limited resources FDA has continually sought to obtain "maximum compliance for minimum dollars expended."[22] FDA admits, for example, that minor offenses involving misbranding have not always been prosecuted. For many of these reasons the FD&C Act, when it was passed in 1938, added injunction to the other legal actions available to FDA.

Before FDA can institute criminal proceedings, informal hearings must be held to determine the merits of the charges pursuant to Section 305 of the FD&C Act. While the agency has only limited access to a company's records, it does have subpoena power to call forth witnesses to testify. (Section 307) Thus, FDA inspectors are instructed to make a special effort to identify by name and title as many key individuals as possible during plant inspections. If an FDA inspector suspects that a plant is not in compliance with good manufacturing practice, he must document the violation in sufficient detail to indicate how and to what extent the process is out of control. He must show a cause and effect relation between the observed deviation and the product quality. Unless the inspector can illustrate the means by which the plant conditions will have an impact on the finished product he will have no grounds for objection.[23]

Additional leverage is given to FDA's law enforcement efforts through two indirect but extremely effective measures. Section 705 of the FD&C Act provides for the dissemination of publicity surrounding any infringement. When fanned by the press media, adverse publicity probably has a more lasting effect on food processors than any legal action. The other recourse is product recall. It is becoming of such widespread utility that it is the subject of an entire chapter of this book.

ENVIRONMENTAL INSPECTIONS

With the public in a state of mass frenzy over ecology, Congress passed a series of sweeping environmental reforms in the 1970's. In rapid succession nine historic bills were enacted: Occupational Safety and Health Act (1970); Clean Air Act (1970); Consumer Product Safety Act (1972); Federal Water Pollution Control Act (amended in 1972); Federal Insecticide, Fungicide and Rodenticide Act (amended in 1972); Safe Drinking Water Act (1974); Hazardous Materials

Transportation Act (1975); Toxic Substances Control Act (1976); and Resource Conservation and Recovery Act (1976). Never before had such comprehensive legislation been passed in so short an interval. For the first time, strict standards were imposed over every facet of manufacturing in the food, chemical, and allied industries. These new laws not only had a shattering impact on industry, but government agencies at times found them all but impossible to implement.

Interagency Regulatory Liaison Group

Reacting to the avalanche of environmental legislation, four federal agencies united in 1977 to form the Interagency Regulatory Liaison Group (IRLG). These agencies, the Food and Drug Administration (FDA), Environmental Protection Agency (EPA), Occupational Safety and Health Administration (OSHA), and Consumer Product Safety Commission (CPSC) were joined in 1979 by a fifth member, the U.S. Department of Agriculture (USDA). Immediately upon its inception IRLG undertook an ambitious program aimed at sharing information, avoiding duplication of effort, and developing consistent regulatory policies.[24] Three projects were noteworthy: instituting coordinated plant inspections, developing criteria for carcinogenicity, and defining suitable agency-company dialogue. The latter effort resulted in a code of conduct (a) that required an outside organization to submit an agenda before it approached regulatory agency personnel; (b) that allowed government staffers discretion in responding to industry questions about impending regulatory action; and (c) that specified minutes be taken at all industry-agency meetings. Henceforth "drop-in" visits by industry representatives would be barred.[25]

A uniform cancer policy among regulatory agencies was badly needed. Back in 1974 when EPA decided to suspend all major uses of the pesticides, aldrin and dieldrin, it applied a definition to carcinogenicity that included any tumor-causing substance. This approach contrasted with FDA's definition which covered only those substances that actually caused cancerous tumors.[26] For many years the agencies have had divergent procedures for conducting animal feeding studies to screen potential carcinogens. Although cancer is as much a mystery as ever, progress has been made in reconciling different government policies. Guidelines issued in 1984 support the significance of animal studies, repudiate the concept of threshold values, and accept the validity of genetic research.[27]

Coordinated plant inspections received priority by IRLG. Three types of inspections were planned: (1) referral inspections where an inspector from one agency would alert a sister agency of any violation, (2) joint inspections conducted by a team of inspectors from different agencies, and (3) "cross-over" inspections made by one official accredited by several agencies. Lacking a cadre of trained inspectors, IRLG never progressed beyond the simplest program, namely, referral inspections. The rationale for these inspections was the "plain view doctrine," which holds that during the normal course of an inspection an official is not required to ignore irregularities which are within sight, even though these problems do not come under his jurisdiction. This doctrine is not a license to pry into areas not under the authority of the inspector's agency.[28]

However well-meaning IRLG may have been, it began to receive severe criticism before it had a chance to prove itself. Senator Talmadge, Chairman of the Committee on Agriculture, Nutrition, and Forestry, stated in 1980 that "such activity on the part of these individual agencies exceeds the mandate the Congress has given them." He went on to allege that the agencies' referral inspection program "puts their inspectors in the role of federal police."[29] Subsequently FDA Commissioner Arthur H. Hayes, Jr. warned that IRLG should not decrease the independence of the member agencies. Culminating this controversy the White House disbanded IRLG in 1981 and assigned its functions to the Office of Science and Technology Policy.[30]

U.S. Department of Agriculture

USDA probably had the least to gain from membership in IRLG. This is because the Federal Meat Inspection Act (FMIA) mandated continuous inspection by USDA. For many years the agency sought to change this constraint as being impractical under modern processing conditions. Its efforts, with support from industry and consumer groups, culminated in major reform at the end of 1986. (See Chapter 4) There has been some debate whether the wording of the Poultry Products Inspection Act, as it now stands, mandates continuous inspection. Doubting that it does, an official of USDA indicated in 1986 that the department planned to take unilateral action to initiate reduced inspection of poultry processors.[31] Unlike the FD&C Act, FMIA explicitly permits the government to inspect and copy all records in connection with the processing of meat products. Plants not in compliance with the regulations may be withdrawn from inspection service after formal proceedings are completed.

Occupational Safety and Health Administration

OSHA, another participant in IRLG, early in its existence devoted its attention to workers' safety by promulgating an endless number of standards. Some 1,100 of these safety regulations eventually proved to be so costly and ineffectual that they were retracted. OSHA's early preoccupation with workers' safety, (e.g., installation of guards over rotating equipment) later gave way to a crusade for workers' health (e.g., reduction of toxic fumes).[32] In its zeal to improve working conditions, the agency asserted its vested powers to conduct warrantless inspections. Furthermore, these inspections, by law, were to be conducted without giving notice.

OSHA's authority to conduct warrantless inspections was challenged by a Pocatello, Idaho, electrical and plumbing contractor, Barlow's, Inc. In a landmark decision the U.S. Supreme Court in 1978 ruled that entries by OSHA without warrants were unconstitutional. Nevertheless the court did recognize the agency's need to conduct inspections. Under the ruling, OSHA could seek an ex parte warrant by showing "probable cause" that OSHA type violations are prevalent in the industry and that a proposed inspection of a specific plant was part of a general enforcement plan.[33]

The Barlow's case prohibiting warrantless searches had repercussions on the conduct of other agencies. EPA and CPSC took immediate steps to comply with this decision affecting factory inspections. In contrast, FDA and USDA were not touched by this ruling because of the unique trust granted them to protect the safety of the food supply. OSHA, although shaken by the Barlow's case, reacted by declaring it would have a minimum impact on its operations. Eula Bingham, Assistant Secretary of Labor for OSHA, gamely advised her over-extended staff "to concentrate on the whales, ignore the minnows." By "whales" she meant significant problems, such as precariously stacked pallets, and by "minnows," technical violations of the standards with no hazards of any consequence.[34] OSHA's lack of clearly defined priorities has left American industry confused. The Chamber of Commerce of the United States has prepared a brochure, *What to Do about OSHA*, in an attempt to steer bewildered manufacturers through the maze of regulations.[35]

Environmental Protection Agency

EPA has the responsibility for enforcing more environmental legislation than any other agency. Among this legislation are two key statutes, the Clean Air Act and the Federal Water Pollution Control Act. Congress intended that both of these acts be "technology forcing." By this term it meant that, in the absence of suitable technology, industry should be encouraged to develop improved methods of pollution control. These new techniques, once perfected, would become the standards against which performance would be measured.[36] The logic of this strategy seemed sound, but it has placed EPA in a potentially compromising position of snooping out the "best available" or "best practical" technology, some of which may be proprietary.

Taking its directives in earnest, EPA unquestionably conducts plant inspections not only to censure transgressors but also to seek out the latest methods of pollution control. Sometimes these actions have overstepped the bounds of accepted conduct. On February 7, 1978, an aerial surveillance team commissioned by EPA flew over the Midland, Michigan, plant of The Dow Chemical Company and snapped detailed photographs of processing equipment. Rather than seek a court warrant to take pictures, an EPA official had sanctioned the flyover after the company had refused to permit cameras in its plant. In every other respect, however, Dow had fully acquiesced to the agency's wishes.[37]

Dow contested this case in court and eventually lost in a 5 to 4 Supreme Court decision which was handed down on May 19, 1986. The court ruled that government investigators do not need warrants to conduct aerial surveillance of areas that any pilot could legally fly over, including fenced yards of private homes and highly secured industrial complexes.[38] Citing this decision, FDA hastily revised its Inspection Operations Manual to instruct its inspectors that they can take photographs during normal inspections without requesting permission from plant management. FDA supported its action by quoting the Supreme Court as saying, "an agency's general inspectional authority need not name every mode of inspection and that photographs and aerial observations are reasonable modes of inspection."[39]

In a second incident involving EPA, a squad of inspectors one day in 1980 arrived at a plant that produces a raw material for the food industry. They were refused admission on the grounds that the group included private consultants. Without safeguards to protect its trade secrets, the company was reluctant to open its doors to a general inspection. The case, already decided in favor of the manufacturer by lower courts, is slated to be heard in the Supreme Court.[40]

EPA administers one of the most comprehensive environmental laws on the books, the Toxic Substances Control Act (TSCA). This statute regulates all synthetic substances not already covered by other legislation; however, there is some overlap in its jurisdiction. If, for example, a manufactured food ingredient, which is regulated under the FD&C Act, is used in a non-food application, it then must comply with the provisions of TSCA. Before a new chemical may be manufactured in commercial quantities or before an existing chemical may be used in new applications a Premanufacture Notification (PMN) must be filed with EPA. Section 6 of TSCA also provides for the regulation of any commercially produced substance that presents an unreasonable risk of injury to health or the environment. A notorious example of such a substance is polychlorinated biphenyls. With TSCA in place only a few years, there is already discussion in Congress about tightening its provisions.[41]

Consumer Product Safety Commission

CPSC, in a Memorandum of Understanding signed with FDA, has assumed jurisdiction over all food containers, such as aerosol cans, that present mechanical risks of injury. On the other hand, the regulation of plastic containers and paper cartons, from which there is a potential migration of substances into the food, resides with FDA.[42] Aerosol propellants was the topic which originally brought CPSC, FDA, and EPA together in discussions that led to the formation of IRLG.[43] Propellants made from chlorofluorocarbons were suspected of depleting the earth's ozone layer in the stratosphere. This shield blocks a significant amount of incident ultraviolet radiation which is damaging to living tissue and may cause skin cancer. As a result of these concerns, the use of chlorofluorocarbon propellants has been severely restricted, and food products in which they are still allowed must have warning labels advising of this environmental danger. (21CFR101.17[c]).

REFERENCES

1. "FDA Warrantless Inspection Court Decision Left Standing," *Food Chemical News*, March 29, 1982, p. 28.
2. "Formal Regulations Governing Inspections Proposed by Neely," *Food Chemical News*, July 3, 1978, pp. 40-43.
3. "Agency Describes Its Surveillance Activities," *Food Chemical News*, May 8, 1978, p. 22.
4. "Self-Incrimination Warning Rules Not Required in FDA Inspections," *Food Chemical News*, January 16, 1978, pp. 37-38.
5. Roger D. Middlekauff, "200 Years of U.S. Food Laws: A Gordian Knot," *Food Technology*, June, 1976, pp. 50, 52.
6. Donald C. Healton, "What's New — A Discussion of Some New FDA Inspectional Techniques," a speech presented to the Annual AFDO Conference, Portland, Oregon, June 23, 1977.
7. Donald C. Healton, a speech presented at the National Association of Pharmaceutical Manufacturers Meeting, Fort Lauderdale, Florida, January 24, 1973.
8. Harold Hopkins, "Consumer Protection: The State Connection," *FDA Consumer*, October, 1978, pp. 14-19.
9. Robert J. Swientek, "FDA Plant Inspection Authority," *Food Processing*, February, 1981, pp 34-36.
10. "What to Do When an FDA Inspector Arrives at Your Plant," Copesan Services, Inc., Milwaukee, Wisconsin, 1981.
11. "Need for Inspection Warrants Denied by FDA Lawyer," *Food Chemical News*, November 20, 1978, pp. 27-29.
12. *The Gold Sheet*, February, 1977, R. P. Scherer Corp., Detroit, Michigan.
13. "Burditt Advises Industry Not to Get Shot in the Act," *Food Chemical News*, August 15, 1977, p. 34.
14. Mary T. O'Brien, "Weighing the Risks and Responsibilities of a HACCP Inspection," *Food Product Development*, December, 1975, pp. 16, 18.
15. Food and Drug Administration, *Field Management Directive* No. 120, January 19, 1979.
16. "FDA Planning to Refuse Studies from Lab Which Repeatedly Declined Inspection," *Food Chemical News*, May 7, 1979, p. 39.
17. Food and Drug Administration, "EDRO Voluntary Correction Reporting System," August 18, 1981.
18. "FDA Proposes Separate 'Regulatory,' 'Adverse Findings' Letters," *Food Chemical News*, June 26, 1978, pp. 56-61.
19. "Notices of Adverse Findings Will Not Be Made Public by FDA," *Food Chemical News*, September 22, 1980, pp. 15-18.
20. "FDA Proposes Separate 'Regulatory, ..., op.cit.
21. Food and Drug Administration, "EDRO Voluntary Correction . . . ., op. cit.

22. "More Use of Injunctions, Prosecutions in Labeling Cases Suggested," *Food Chemical News*, October 31, 1977, pp. 29, 30.

23. Food and Drug Administration, "Agency Strategy for Completion of the Investigation at _____," Intra-Agency Memorandum, June 2, 1977.

24. "Interagency Regulatory Liaison Group," *Federal Register*, May 22, 1979, pp. 29822-29825.

25. *Chemical Engineering*, January 2, 1978, pp. 19, 20.

26. "EPA Broadens Its Definition of Carcinogen," *Chemical and Engineering News*, October 14, 1974, p. 13.

27. "Guides Proposed on Cancer Agents," *The New York Times*, May 23, 1984, p. B28.

28. "Referral Inspections and the IRLG," *Food Processing*, February, 1981, p. 35.

29. "NFPA Says IRLG Has Changed Its Position on Notice of Referral Inspection," *Food Chemical News*, July 7, 1980, pp. 14, 15.

30. "IRLG Folded under White House Wings," *Food Chemical News*, October 5, 1981, p. 3.

31. "Illegal Activity in Poultry Industry Alleged by U.S. Attorney," *Food Chemical News*, September 22, 1986, pp. 36-38.

32. Tom Alexander, "OSHA's Ill-Conceived Crusade Against Cancer," *Fortune*, July 3, 1978, pp. 86-90.

33. "OSHA Inspections," *Food, Drug & Cosmetic Manufacturing*, July, 1978.

34. "Key to OSHA Inspections: 'Don't Become Uptight'," *Package Engineering*, November, 1980, p. 15.

35. Chamber of Commerce of the United States, *What to Do about OSHA*, Washington, D.C., 1978.

36. "The Clean Water Act," *Processed Prepared Food*, June, 1979, pp. 25-39.

37. Edward M. Nussbaum, Garry L. Hamlin, "Aerial Surveillance of Plants: Is It Ethical?" *Chemical Engineering Progress*, April, 1981, pp. 15, 16.

38. "Aerial Searches of Fenced Areas Upheld by Court," *The New York Times*, May 20, 1986, pp. A1, A20.

39. "FDA Denies Policy on Photos during Inspections Has Changed," *Food Chemical News*, October 27, 1986, pp. 18, 19.

40. "Can a Company Lock Out the EPA's Inspectors?" *Business Week*, January 9, 1984, p. 122.

41. "Changes in New Chemical Rules Drafted," *Chemical and Engineering News*, July 9, 1984, p. 18.

42. "FDA and CPSC Sign Memo of Understanding," *Food Technology*, October, 1976, p. 121.

43. "One for All Four," *Chemical Week*, August 10, 1977, p. 18.

CHAPTER SEVEN
KOSHER CERTIFICATION

Kosher foods are defined as products prepared according to Jewish dietary laws and under the supervision of recognized authorities. Dating back to antiquity, kosher practices are undoubtedly the oldest code of food laws now in existence. These practices are grounded in Hebrew dogma which antedates the Christian era. The embodiment of the kosher dietary laws is known as the Kashruth. Although these rules are sometimes likened to food sanitary standards, the principal motivation for adhering to them is religious.

KASHRUTH LAW

The fundamental teachings of the Kashruth are a remarkable set of passages that span human nutritional and spiritual needs. These writings are scattered throughout the Torah, which is known as the Five Books of Moses, consisting of Genesis, Exodus, Leviticus, Numbers, and Deuteronomy.[1] Christians recognize these books as the Pentateuch of the Bible. While the basic tenets of Kashruth have been amplified and embellished to cover situations not encountered in ancient Israel, the original scriptures are timeless. A partial summary of the relevant verses includes:

Gen.1.29   God said, "See, I give you every seed-bearing plant that is upon all the earth, and every tree that has seed-bearing fruit; they shall be yours for food."

9.3   Every creature that lives shall be yours to eat; as with the green grasses, I give you all these.

9.20-21   Noah, the tiller of the soil, was the first to plant a vineyard. He drank of the wine and became drunk. . . .

32.33   That is why the children of Israel to this day do not eat the thigh muscle that is on the socket of the hip, since Jacob's hip socket was wrenched at the thigh muscle.

171

Ex. 13.3 And Moses said to the people, "Remember this day, on which you went free from Egypt, the house of bondage, how the Lord freed you from it with a mighty hand: no leavened bread shall be eaten."

16.24 So they put it aside until morning, as Moses had ordered; and it did not turn foul, and there were no maggots in it.

23.19 . . . you shall not boil a kid in its mother's milk.

Lev. 2.13 You shall season your every offering of meal with salt; you shall not omit from your meal offering the salt of your covenant with God; with all your offerings you must offer salt.

7.23 Speak to the Israelite people thus: You shall eat no fat of ox or sheep or goat.

7.26 And you must not consume any blood, either of bird or of animal, in any of your settlements.

11.3 Any animal that has true hoofs, with clefts through the hoofs, and that chews the cud — such you may eat.

11.20 All winged swarming things [flying insects], that walk on fours, shall be an abomination to you.

11.29 The following shall be unclean for you from among the things that swarm on the earth: the mole, the mouse, and great lizards of every variety.

11.32 And anything on which one of them falls when dead shall be unclean: be it any article of wood, or a cloth, or a skin, or a sack — any such article that can be put to use shall be dipped in water, and it shall remain unclean until evening; then it shall be clean.

11.39 If an animal that you may eat has died, anyone who touches its carcass shall be unclean until evening.

11.42 You shall not eat, among all things that swarm upon the earth [crawling insects], anything that crawls on its belly, or anything that walks on fours, or anything that has many legs; for they are an abomination.

19.23 When you enter the land and plant any tree for food, you shall regard its fruit as forbidden. Three years it shall be forbidden for you, not to be eaten.

Deut. 14.9 These you may eat of all that live in water: you may eat anything that has fins and scales.

14.11 You may eat any clean bird.

22.6 If, along the road, you chance upon a bird's nest, in any tree or on the ground, with fledglings or eggs and the mother sitting over the fledglings or on the eggs, do not take the mother together with her young.

Interpretations of Food Laws

Interpretations of the food laws as given in the Torah are complex and only thoroughly understood by qualified persons. Nevertheless, certain general guidelines can be outlined for the food manufacturer. Perhaps the simplest way of introducing the subject to laymen is to divide all food into three broad categories, animal, vegetable, and mineral. The kosher rules for animals are the most fastidious and encompass detailed instructions about what can be eaten and how it is to be prepared. All fruits and vegetables are accepted as inherently kosher and are of interest primarily to ensure their proper processing with other kosher ingredients. Other than salt, no mention is made in the Torah of mineral products. Thus, synthetic compounds derived from inorganic chemicals or petroleum substances are outside the concerns of kosher rules except as they might become contaminated with non-kosher materials.[2] More specifics about kosher foods are given in the following groupings.

Meat, according to the dictum that acceptable animals must have true hoofs with clefts through the hoofs and chew the cud, may be eaten if it is beef or mutton. On the other hand, pork is strictly taboo. So too is horse and rabbit meat. All kosher animals must be slaughtered in a ritualized manner and dressed so as to drain all the blood and trim the hard fat or suet. The hindquarters are not permitted unless meticulous care is taken to remove the sciatic nerve in the thigh. Diseased and imperfect animals must be discarded.

Poultry for consumption includes as a rule only barnyard fowl. Birds of prey and carrion eaters are proscribed. Poultry must be carefully drained of all blood. As is done with meat, poultry is koshered with coarse salt to draw out the last vestiges of blood before cooking.

Fish is restricted to those species that have fins and scales. The cataloguing of acceptable fish is not completely settled and is still open to question. Tuna fish, after some debate, has been included in the approved list.[3] Clearly forbidden are all shellfish, both molluscan and crustacean.

Eggs may be eaten provided they are taken from kosher birds. The only reference to eggs in the Torah is from Deuteronomy 22.6 so that most of the rulings about eggs have been arrived at by inference.[4] All eggs should be inspected to insure that they are free from blood spots. Any blemished eggs must be discarded.

Milk is required to come from kosher animals. Even though they may be kosher, milk and dairy products cannot be consumed at the same meal at which meat is eaten. (cf. Ex. 23.19) To avoid the possibility that dairy products and meat might be co-mixed, they must be prepared separately. Foods that may be eaten with either dairy or meat dishes are called pareve. Such foods as fruits, vegetables, fish, and eggs are considered to be pareve. Cheeseburgers are totally forbidden, and ice cream is not permitted when meat is served. An ice cream substitute called tofu, based on a soy protein and with butterfat being replaced by vegetable oils, is being marketed as pareve.[5]

Vegetables and fruits generally do not require rabbinical supervision. Certain ones, however, such as tomatoes and beans are commonly processed in plants that also handle non-kosher foods such as pork. Thus, these particular products need to be certified as being kosher. All fruits and vegetables should be examined to make sure they are free of insects. Fruits picked from trees less than three years old are non-kosher.

Wine and grape products, including grape juice, wine vinegar, and brandy, are in their own special class. These products must be produced and handled completely by observant Jews in order to be acceptable as kosher. This rule is not based on any food law found in the Torah but on the exhortation against fraternizing with non-Jews in order to discourage intermarriages. (cf. Deut. 7.3) Once grape products are pasteurized, though, they may be distributed through normal marketing channels.

Allowances or Annulments

Taken at face value the kosher rules are unbending and therefore difficult to apply to modern food technology. In practice, however, allowances or annulments may be made for special circumstances. Thus if an ingredient undergoes such extensive chemical and physical processing that an intermediate is unfit to eat, then the identity of the source of the ingredient may be disregarded. In support of this position, an ancient Jewish law has been quoted:

> The skin of the stomach is sometimes salted and dried so that it is rendered like wood, and if then it is filled with milk it is kosher, because after the stomach has been dried, it is just like wood, and does not contain any moisture of meat.[6]

Another plausible occurrence is when kosher food becomes contaminated with a non-kosher substance. In these instances a degree of reasonableness prevails. For example, if a drop of milk falls into a meat stew more than sixty times its volume, the dish may still be eaten.[7] And again should olive oil or wine produced from trees or vines less than three years old contaminate the same kinds of products that are kosher, the whole must be rejected unless the fraction of good produce is 200 times the volume of the contaminant.[8]

The above principles are useful in determining the acceptability of certain food ingredients although rabbinical authorities are not unanimous on some points. Because wine argols are subjected to lengthy chemical processing, the resulting cream of tartar is accepted outright.[9] Vitamins used for food fortification are similarly permitted even though certain non-kosher fish oils may be used as raw materials.[10] Microbiologically produced rennin is approved, but rennin extracted from animal stomachs may be disallowed unless the animal is kosher. Since active rennin cannot be detected in modified whey products which have been subjected to extensive chemical and heat treatments, these products are generally classified as kosher for dairy use. Another ingredient considered kosher for dairy use is sodium caseinate notwithstanding its use in so-called "non dairy" creamers. All food emulsifiers must be produced from vegetable oils instead of animal fats. For the same reason only glycerine derived from petroleum sources is permitted.

No food ingredient is more controversial than gelatin. This colloid is an ideal functional ingredient for such products as frozen desserts, but because an economical supply of gelatin from kosher-slaughtered animals is not available in the United States, its use has been severely restricted. Made from animal hides and bones, whatever kosher gelatin is available is considered to be pareve. Although no manufacturer of gelatin from non-kosher animals has been certified by a rabbinical association, a number of prepared gelatin desserts have been approved. In this special case, discretion has been given to the consumer to decide in good conscience what is correct. As changes in gelatin manufacturing are accepted, the status of this ingredient could be reevaluated. Newer processes based on ossein could qualify as kosher once all the facts have been reviewed.

Considerations of Health

Even though kosher laws have their main appeal as a religious doctrine, they have proven to be scientifically sound. Abstinence from eating pork and rabbit meat has spared generations of Jews from being afflicted with trichinosis and turlaremia. Since germs of many diseases are carried in the bloodstream, the prohibition against eating blood is also sound. Even forbearance from eating meat and dairy products together is dietetically reasonable. The unequal rates of digestion of meat and milk proteins suggest that these foods be consumed separately.[11] With a few prominent exceptions many of the Jewish dietary laws have been assimilated into our Western heritage. Of particular interest to the food manufacturer is the fact that Kashruth in no way conflicts with the GMP's.

FACTORY CERTIFICATION

The incentive on the part of food manufacturers to obtain kosher certification is the fact that about three percent of the United States population is Jewish. Furthermore, demographic studies indicate that this minority is concentrated in such population centers as New York, Boston, and Los Angeles so that its influence exceeds what would be expected from the national average. Admittedly not all Jews are strict followers of the Kashruth; of the three major sects, Orthodox, Conservative, and Reform, only the Orthodox Jews are sticklers for rigid obedience to the food laws. Other religious groups, however, demand kosher food. Moslems, and among Christians, the Seventh-Day Adventists and Seventh-Day Baptists, look to these foods for religious fulfillment.

Reflecting the growing interest in kosher foods, the first-ever Kosher Food and Life Expo was held in March, 1987, at the Jacob Javits Convention Center in New York City. With a gate of more than 50,000 people, the event proved to be a huge success. The reception to this trade show can be explained by the fact that there are over 10,000 kosher food products available nationwide, up from fewer than 4,000 a decade ago. The value put on these products is more than $1 billion,with a potential four times as great.[12]

Given these statistics, many of the major food companies have recognized the need to obtain kosher certification for their plants. Distribution costs being as great as they are, wholesalers and retailers are not able to carry two complete sets of stocks, one kosher and the other non-kosher. In turn, food manufacturers increasingly require that their suppliers furnish only kosher grade ingredients and materials. Aside from the consideration of inventory control, certified plants must keep their equipment scrupulously free of all non-kosher ingredients. To illustrate, a manufacturer which wants to produce both kosher and non-kosher emulsifiers in the same reactor must thoroughly clean out its equipment before starting a batch of kosher product. This changeover requires that the reactor be flushed out with vegetable oil or preferably boiling water and then left idle for 24 hours.[13] Under such circumstances most producers decide that it is easier to produce only kosher products.

Rabbinical Associations

Plant certification can be requested from any one of a number of nonprofit rabbinical associations. Unlike some religious orders, there is no hierarchical Jewish organization. All of the several rabbinical services have equal status and therefore may be used interchangeably. Most of the major urban centers have local rabbinical associations which in some cases provide nationwide services. The largest and one of the most respected is the Union of Orthodox Jewish Congregations of America with headquarters in New York. It began its operations in 1925 and by 1961 was supervising 1830 products for 359 companies.[14] Table 7.1 lists several other established services with fine reputations.[15] In selecting a certification service, a food manufacturer should consider only Orthodox associations inasmuch as their interpretations of the law are regarded as most stringent.

An application for certification is generally made by submitting a form giving details of the food process including the sources of all ingredients and supplies. This application is followed by a plant inspection. If the request is accepted, a contract is signed specifying the fee and the conditions of approval. The annual cost may vary widely from plant to plant depending on needs, but the range for typical services was mentioned as being $750 to $1200 in 1979.[16] Obviously if continuous inspection is required, the figure can run much higher. The frequency of inspection may vary from continuous supervision to daily, weekly, monthly, or quarterly visits. Not an inconsequential benefit of certification is the right to

TABLE 7.1
RABBINICAL ORGANIZATIONS

NAME	LOCATION	LOGO
Ko Kosher Service	Philadelphia, PA	**Ko**
Kosher Overseers Association of America, Inc.	Beverly Hills, CA	**K**
Kosher Supervision Service of New Jersey	Teaneck, NJ	🄰
Organized Kashruth Laboratories	Brooklyn, NY	Ⓚ
Union of Orthodox Jewish Congregations of America	New York, NY	Ⓤ
Vaad Harabonim	Boston, MA	🄰

use any such logo or trademark owned by the rabbinical association on labels, in promotional material, and in advertising. Certified foods frequently are listed free of charge in Jewish organs and product directories.[17]

Besides food ingredients, all supplies used by a certified plant must be kosher. Packaging, for example, needs to approved. Plastics, such as polyethylene, which are produced from petrochemicals are acceptable. Glycerine, widely used in cellophane manufacture, must be based on petroleum. All detergents, defoamers, and lubricants have to be synthesized from non-animal raw materials. As long as kosher products are tightly sealed in shipping containers they may be stored with non-kosher materials of similar nature. Bulk shipments, however, require cars and trucks dedicated to single use in order to avoid tedious cleaning and inspection procedures. Overall, the cautious policy to follow is to keep all non-kosher ingredients and supplies out of a certified plant, thereby avoiding accidental contamination.

CONTINUING KOSHER GUARANTY

The labeling of kosher products is regulated by federal and state laws. Federal regulations require that:

> The term "kosher" should be used only on food products that meet certain religious dietary requirements. The precise significance of the phrase "kosher style" as applied to any particular product by the public has not been determined. There is a likelihood that the use of the term may cause the prospective purchaser to think that the product is "kosher." Accordingly, the Food and Drug Administration believes that use of the phrase should be discouraged on products that do not meet the religious dietary requirements. (21CFR101.29)

State Laws

New York State law declares that all foods represented as kosher must be prepared in accordance with Orthodox Hebrew religious requirements. Otherwise a misdemeanor has been committed punishable by a maximum penalty of $1000 and a year in jail.[18,19] Recent legislation requires that stores must indicate whether they exclusively sell kosher items or if both kosher and non-kosher foods are carried. Manufacturers who wish to market kosher-labeled products in New York State must file with the Department of Agriculture and Markets the name of the rabbi responsible for certification. New Jersey has passed similar legislation. Attempts to control kosher labeling, however, are confounded by the imprecise use of such terms as "kosher pickles" and "Jewish rye bread."[20]

Certified kosher products display the logo of the supervising rabbinical association, or if one is not available then they are marked with a "K." The letter symbol is generic for all kosher-approved products and therefore not copyrighted. Because of the variety of logos in use, some experience is required to become familiar with them. The marks used by some of the better known associations are shown in Table 7.1. Although a food ingredient or product may meet all the requirements of the Kashruth, it is not allowed to be marked kosher unless it is certified by a recognized rabbinical association. Often manufacturers desire to obtain certification for products like table salt that do not require rabbinical supervision. They do this purposely so as to be able to label their products kosher and thereby gain a marketing advantage.

Kosher for Passover

Foods eaten at Passover must conform to special dietary laws based on the admonition in Exodus 13.3 not to eat leavened bread. This prohibition extends to cakes, biscuits, crackers, cereals, wheat, barley, oats, rice, dry peas, and dry beans. Also forbidden during the holy observance are all liquids containing flavors or ingredients from grain alcohol. All approved foods must be prepared under the supervision of a rabbi, be labeled "kosher for Passover," and bear a rabbinical signature or certificate.[21]

Food processors regularly require letters of guaranty from their suppliers stating that purchased ingredients and materials have been produced under the supervision of a rabbi. For those products such as baking soda not requiring rabbinical supervision, an advisory letter to this effect usually can be obtained from a rabbinical association and sent to customers on request. Since supervisory contracts invariably are written on an annual basis with automatic renewal clauses, letters of certification need to be updated yearly. For a large food manufacturer the paper work can easily become significant. Therefore several companies have proposed the implementation of a continuing guaranty form such as illustrated in Figure 7.1. Of course, such a document would be backed up by rabbinical letters on file.

To: _____

       _____

       _____

Gentlemen:

       _____ hereby guaranties, that, as of the date of each shipment by it to you of any product which is labeled kosher, such product shall have been prepared, manufactured, and packaged in strict accordance with the current procedures prescribed by an accepted Jewish rabbinical authority for kosher certification; and that the given product, when shipped, shall be totally free from contamination so as to be in compliance with the standards of quality and purity required for displaying "K", or other recognized kosher symbol.

       This guaranty shall be a continuing guaranty, and it shall continue in effect until such date as you shall receive written notice of the revocation of the guaranty contained herein.

DATE: _____   BY: _____

FIGURE 7.1. CONTINUING KOSHER GUARANTY

REFERENCES

1. *The Torah*, The Jewish Publication Society of America, Philadelphia, 1962.
2. Seymour E. Freedman, *The Book of Kashruth*, Block Publishing Company, New York, 1970, pp. 149, 253-262.
3. Ibid, p. 129.
4. John W. Ellison, *Nelson's Complete Concordance of the Revised Standard Version Bible*, Thomas Nelson & Sons, New York, 1957.
5. "It's Trendy, Tasty and Tofutti," *Time*, July 9, 1984, p. 18.
6. "What's Needed for Kosher Approval," *Food Engineering*, August, 1978, pp. 87, 88.
7. Morris Golomb, *Know Jewish Living and Enjoy It*, Shengold Publishers, Inc., New York, 1981, p. 102.

8. Dayan Dr I. Grunfeld, *The Jewish Dietary Laws*, Vol. 2, The Soncino Press, London, 1972, p. 35.

9. *Kashruth Handbook for Home and School*, Union of Orthodox Jewish Congregations of America, New York, 1972.

10. Joe M. Regenstein and Carrie E. Regenstein, "An Introduction to the Kosher Dietary Laws for Food Scientists and Food Processors," *Food Technology*, January, 1979, pp. 89-99.

11. Morris Golomb, op. cit., pp. 89-106.

12. David Tuller, "What's New in Kosher Food," *The New York Times*, April 5, 1987, p. F23.

13. David Druckman, "The Quality of Kosher Foods," *Quality Progress*, June, 1977, pp. 16-18.

14. *The Key to the Kosher Market*, Union or Orthodox Jewish Congregations of America, New York, 1961.

15. "Kosher Products Guide," *Kosher Home*, October, 1978, pp. 38, 43.

16. Israel Shenker, "With Them, It's Always Strictly Kosher," *The New York Times Magazine*, April 15, 1979, pp. 32-42.

17. *Industrial & Institutional Directory of Kosher Products & Services*, Union of Orthodox Jewish Congregations of America, New York, 1982.

18. Leonard Sloane, "Calling It Kosher: How to and Why," *The New York Times*, May 18, 1975.

19. State of New York, Department of Agriculture and Markets, *Circular 811*, Albany, New York 12235, Revised June, 1986.

20. "Jersey Tightens Up on Kosher Labels," *The New York Times*, December 28, 1983, p. C14.

21. Samuel H. Dresner and Seymour Siegel, *The Jewish Dietary Laws*, The Burning Bush Press, New York, 1959, pp. 63, 64.

PART III

Customer Service

CHAPTER EIGHT
FDA GUARANTY LETTER

Customer service in the food industry requires a major commitment from the Quality Assurance Department (QA) since it is the only corporate group equipped to handle the many intricacies of this service function. Among the numerous demands made by customers are requests for FDA (Food and Drug Administration) guaranty letters and for technical data sheets. QA also is given the primary responsibility to investigate product complaints and resolve them to the satisfaction of customers. These customer service activities are reviewed in this and the next two chapters. The discharge of these duties often taxes the full patience and ingenuity of quality assurance personnel. Successful customer service depends on providing reliable answers and exercising not a little diplomacy.

LEGAL PROVISIONS

An FDA guaranty letter is a written guaranty submitted to a customer stating that the products covered by the guaranty are in compliance with the requirements of the Federal Food, Drug, and Cosmetic Act (FD&C Act). Section 303(c)(2) and (3) of the Act states:

(1) No person shall be subject to the penalties of subsection (a) of this section . . .
(2) for having violated Section 301(a) or (d), if he establishes a guaranty or undertaking signed by, and containing the name and address of, the person residing in the United States from whom he received in good faith the article, to the effect . . . that such article is not adulterated or misbranded . . . or . . . is not an article which may not, under the provisions of Section 404 or 505, be introduced into interstate commerce; or
(3) for having violated section 301(a), where the violation exists because the article is adulterated by reason of containing a color additive not from a [certified] batch . . . if such person establishes a guaranty or undertaking signed by, and containing the name and address of, the manufacturer of the color additive, to the effect that such color additive was from a [certified] batch. . . .

The above provisions excuse any person from prosecution if he introduces an adulterated or misbranded food product into interstate commerce provided he has in his possession a guaranty from his supplier that such product meets all federal regulations.

Flavoring Materials

A similar guaranty may be issued to cover flavoring materials. Such a provision is specified in 21CFR101.22(i)(4):

> A flavor supplier shall certify, in writing, that any flavor he supplies which is designated as containing no artificial flavor does not, to the best of his knowledge and belief, contain any artificial flavor, and that he has added no artificial flavor to it. The requirement for such certification may be satisfied by a guarantee under Section 303(c)(2) of the act which contains such a specific statement. . . .

USDA Regulations

An additional reason why a customer might request an FDA guaranty letter relates to U.S. Department of Agriculture (USDA) regulations promulgated under the Federal Meat Inspection Act and the Poultry Products Inspection Act. The Manual of Meat Inspection Procedures (Section 302.3) and the Poultry Inspector's Handbook (Section 81.95) both specify that all non-meat ingredients, such as curing agents and spices, which are not produced in USDA inspected plants must be covered by FDA guaranty letters. In 1984 this requirement was extended to include all packaging materials.[1]

At the end of 1986 USDA instituted a policy that requires all suppliers of additives to a plant producing meat or poultry products to submit continuing guaranties that such additives are free of sulfites, or if not, to indicate their maximum levels. For the purpose of these guaranties, the cut-off level below which sulfites can be considered negligible is 10 ppm. These guaranties will be relied upon by USDA to supervise the labeling of sulfites in all products under its jurisdiction. The assumption being made is that whatever sulfites contained in meat and poultry products are introduced exclusively by additives and therefore can be determined by a simple calculation rather than laborious analysis. Guaranty letters and calculations using data from these letters must be kept on file by the plant.[2,3]

State or Municipal Laws

Customers may also request guaranties to seek immunity from legal actions under state or municipal laws which are identical or substantially similar to the FD&C Act. Aside from the special requirement that all meat and poultry plants have guaranties on file, the principal protection afforded by any FDA guaranty letter accrues to such middlemen as food jobbers and distributors. Conversely, the food manufacturer which receives a guaranty for an ingredient still has a general obligation under Current Good Manufacturing Practice to utilize "chemical, microbiological, or extraneous-material testing procedures . . . where necessary to identify sanitation failures or food contamination." (21CFR110.80[g])

FORM LETTERS

To assist the food manufacturer in responding to requests for FDA guaranty letters, a number of standard forms are available. Other letters have been adopted by modifying or expanding these basic forms. The simplest statement is one proposed by USDA for inclusion on an invoice or bill of sale and is shown in Figure 8.1.[4] FDA proposes two guaranty forms, a limited one as well as a general and continuing instrument. The latter is reproduced in Figure 8.2. Also a color additive guaranty is shown in Figure 8.3 (21CFR7.13)

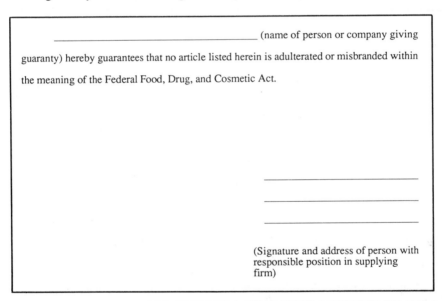

_____ (name of person or company giving

guaranty) hereby guarantees that no article listed herein is adulterated or misbranded within

the meaning of the Federal Food, Drug, and Cosmetic Act.

(Signature and address of person with responsible position in supplying firm)

FIGURE 8.1. LIMITED FORM FDA GUARANTY ACCEPTED BY USDA

Modified Forms

While the forms suggested by USDA and FDA are perfectly adequate to meet the requirements of the regulations, they may fail to address specific needs of the food manufacturer or his customers. For this reason a modified form is shown in Figure 8.4. It substitutes a schedule of products covered by the guaranty for an open-ended category of "articles." Products listed in the schedule may be identified by their name, grade, and if necessary the manufacturing location. The form protects the supplier by including a clause that nullifies the guaranty in the event that a product is not used in accordance with the FD&C Act or state laws.

The article comprising each shipment or other delivery hereafter made by

_____ (name of person giving the guaranty or

undertaking) to, or in the order of

_____ (name and post-office address

of person to whom the guaranty or undertaking is given) is hereby guaranteed, as of the

date of such shipment or delivery, to be, on such date, not adulterated or misbranded

within the meaning of the Federal Food, Drug, and Cosmetic Act, and not an article which

may not, under the provisions of Section 404, 505, or 512 of the Act, be introduced into

interstate commerce.

_____

_____

_____

(Signature and post-office address of

person giving the guaranty or

undertaking.)

FIGURE 8.2. FDA PROPOSED GUARANTY

_____ (Name of manufacturer) hereby

guarantees that all color additives listed herein were manufactured by him and (where color

additive regulations require certification) are from batches certified in accordance with the

applicable regulations promulgated under the Federal Food, Drug, and Cosmetic Act.

_____

_____

_____

(Signature and post-office address of

manufacturer)

FIGURE 8.3. COLOR ADDITIVE GUARANTY UNDER SECTION 303(c)(3)
OF THE FEDERAL FOOD, DRUG, AND COSMETIC ACT

TO: _____

_____

_____

(Name and post-office address of Receiver)

    For the purpose of Section 303(c)(2) of the Federal Food, Drug, and Cosmetic Act (the Act) and for no other purpose _____ (Supplier) hereby guarantees that, as of the date of each shipment by it to you of any article listed in Schedule A hereto, such article is not, when shipped, adulterated or misbranded within the meaning of the Act. The Supplier further guarantees that, as of the date of each such shipment, no such article is an article which may not, under the provision of Section 404 or 505 of the Act be introduced into interstate commerce.

    This guaranty shall, however, be void and of no effect in any instance where the particular use by you or your customer of any article to which this guaranty would otherwise apply is a use which is not in accordance with the requirements of the Act or applicable state laws.

    This guaranty shall continue in effect until such date as you shall receive from the Supplier written notice of the revocation of the guaranty contained herein.

DATE:_____   BY:_____

_____

_____

(Signature and post-office address of Supplier)

SCHEDULE A

FIGURE 8.4. MODIFIED FDA GUARANTY

Finally, the phrase, ''when shipped,'' is inserted into the guaranty in recognition of the fact that after the supplier loses control of the product he no longer can vouch for its integrity.

Experience with using a form similar to the one shown in Figure 8.4 has revealed certain drawbacks with it. Each time a new product is shipped to an existing customer, a new guaranty needs to be prepared. Both the supplier and receiver may have difficulty keeping track of which products have been covered by a guaranty and which ones may have been overlooked. Therefore, to reduce the chance of omission and to cut down on the administrative load, a continuing guaranty of the form illustrated in Figure 8.5 has been devised. It is designed for only those products labeled, ''food grade,'' which of necessity must conform

TO: _____

_____

_____

(Name and post-office address of Receiver)

For the purpose of Section 303(c)(2) of the Federal Food, Drug, and Cosmetic Act (the Act) and for no other purpose, _____ (Supplier) hereby guarantees that, as of the date of each shipment by it to you of any product which is labeled "Food Grade," such product is not, when shipped, adulterated or misbranded within the meaning of the Act. The Supplier further guarantees that, as of the date of each such shipment, no such product is an article which may not, under the provision of Section 404 of the Act, be introduced into interstate commerce. This guaranty shall also apply under substantially identical state or municipal regulations.

This guaranty shall, however, be void and of no effect in any instance where the particular use by you or your customer of any product to which this guaranty would otherwise apply is a use which is not in accordance with the requirements of the Act or applicable state laws.

By the acceptance of this guaranty, you agree to promptly notify the Supplier in writing of any demand, complaint or proceeding within your knowledge for claimed violation of the Act resulting from or in any manner arising out of any such shipment, including the name and address of the complainant and the name of the product involved.

This guaranty shall continue in effect until such date as you shall receive from the Supplier written notice of the revocation of the guaranty contained herein.

DATE: _____   BY: _____

                                      _____

                                      _____

(Signature and post-office address of Supplier)

FIGURE 8.5. CONTINUING GUARANTY UNDER SECTION 303(c)(2) OF THE FEDERAL FOOD, DRUG, AND COSMETIC ACT

to the FD&C Act. The guaranty extends to any situation covered by substantially identical state or municipal ordinances. In addition, a paragraph is included which requires the receiver to notify the supplier of any complaint made with regard to the product shipped. A number of companies incorporate such a provision in their guaranty forms. While the form shown in Figure 8.5 is useful for all "food grade" products, any time a guaranty is provided for animal feed ingredients or bulk pharmaceuticals the previous form given in Figure 8.4 must be used.

An important condition in the FD&C Act is that an FDA guaranty can only be given to an immediate customer or consignee. If a distributor or jobber takes title to the goods, then the ultimate customer must seek a guaranty from that distributor, not from the manufacturer. The distributor in turn, however, may ask for a guaranty from the manufacturer. A broker whose only function is to find a buyer and is paid by a commission would not be affected by these constraints. Section 301(h) of the FD&C Act excuses any person from the penalties of the act when giving a guaranty that is false if he, in good faith, is relying upon a guaranty to the same effect received from his supplier.

Special Forms

There is no particular reason why an FDA guaranty letter cannot be given to an overseas customer. What advantage this customer might gain from obtaining such a guaranty would depend on local law. In some instances a customer might desire a guaranty under a foreign food law. Thus a guaranty as shown in Figure 8.6 might be supplied to a Canadian customer, specifying compliance with the Canadian Food and Drugs Act.

As indicated earlier in this chapter, a guaranty may be issued to cover flavoring materials. Such a guaranty would certify that a product does not contain any substance which is classified as an artificial flavor. A form for this purpose is shown in Figure 8.7. As with any of the other FDA guaranties mentioned above, the making of an erroneous statement with regard to a flavor is prohibited under the FD&C Act and is punishable by fines and/or imprisonment.

SUPPLY CONTRACTS

Along with a request for an FDA guaranty letter, a customer may ask for any number of signed statements. Parts of these agreements may be forwarded for handling to various corporate departments, such as Personnel, Insurance, or Law. But because QA often is the only operational group to see these forms it should know how to coordinate the responses. An FDA guaranty clause frequently is incorporated into a customer's written specification for an ingredient. An FDA guaranty may also be inserted into a general supply contract. Before any form

TO: _____

_____

_____

(Name and address of Receiver)

　　For the purpose of the Canadian Food and Drugs Act, and for no other purpose,

_____ (Supplier) hereby guarantees, that, as of the date of each

shipment by it to you of any article listed in Schedule A hereto, such article is not, when

shipped, adulterated or misbranded within the meaning of Part I, Sections 4-7 of said Act

and/or regulations promulgated thereunder affecting said article.

　　This guaranty shall, however, be void and of no effect in any instance where the

particular use by you or your customer of any article to which this guaranty would

otherwise apply is a use which is not in accordance with the requirements of the Canadian

Food and Drugs Act or regulations promulgated thereunder.

　　This guaranty shall continue in effect until such date as you shall receive from the

Supplier written notice of the revocation of the guaranty contained herein.

DATE: _____　　　　BY: _____

　　　　　　　　　　　　　　　　　_____

　　　　　　　　　　　　　　　　　_____

(Signature and address of Supplier)

SCHEDULE A

FIGURE 8.6. CANADIAN GUARANTY UNDER THE FOOD AND DRUGS ACT

submitted by a customer is signed it should be carefully reviewed, and each section should be considered on its own merits. Requests for Kosher certificates, for example, have already been discussed in Chapter Seven.

Equal Employment Opportunity Affidavits

　　Other conditions which may accompany an FDA guaranty are highlighted below. Government contractors have a duty to abide by fair labor practices and other socially oriented legislation. These requirements extend to subcontractors who in many instances are required to submit signed affidavits of compliance. Thus the Fair Labor Standards Act of 1938 as amended provides for minimum hourly wages and a maximum work week for employees. The Civil Rights Act of 1964 and Executive Order 11246 specify Equal Employment Opportunity (EEO) requirements that prohibit an employer from discriminating against members of

TO: _____

_____

_____

(Name and post-office address of Receiver)

For the purpose of 21CFR101.22 and for no other purpose,

_____ (Supplier) certifies that, as of the date of each shipment

by it to you of any article listed in Schedule A hereto, such article does not, to the best of

the knowledge and belief of the Supplier, when shipped, contain any artificial flavor as

defined in 21CFR101.22(a)(1), and that the Supplier has added no artificial flavor to it.

This certificate shall continue in effect until such date as you shall receive from the

Supplier written notice of the revocation of the certification contained herein.

DATE:_____   BY: _____

_____

_____

(Signature and post-office address of

Supplier)

SCHEDULE A

FIGURE 8.7. CERTIFICATE UNDER 21CFR101.22

its workforce — because of race, religion, color, sex, or national origin.[5] Furthermore, segregated facilities are not permitted at any work location. Public Law 93-112 forbids discrimination against the physically and mentally handicapped worker. Finally, the Small Business Act (Public Law 95-507) gives favored treatment in government contracts to those firms in any one of three classifications: 1. women-owned business, 2. small business, or 3. small disadvantaged business.

Indemnity Clauses

Indemnity clauses sometimes are subtly slipped into FDA guaranty forms. A typical one may read:

The Seller further agrees to indemnify, defend and hold the Buyer harmless from and against any and all liability, loss or expense arising out of or resulting from any claim based on the use or consumption of any article sold by the Seller to the Buyer as long as the Seller is notified of such claim within a reasonable period of time and is permitted to deal therewith in its own discretion and through its own representative or attorney.

Such a clause when executed by the seller may expose him to a substantial liability. Therefore if the seller feels obligated to sign an indemnity clause, instead of including it as part of an FDA guaranty, he would be well advised to incorporate it as a condition of any sales contract entered into between the seller and buyer on a case-by-case basis.

Certificates of Insurance

Certificates of insurance are commonly requested at the time FDA guaranties are obtained. These certificates, issued under the signature of an underwriter, specify that a given amount of product liability insurance is in force. Although not considered a substitute for an FDA guaranty letter, a certificate of insurance in some respects provides more protection to the buyer. Particularly if the seller is small, undercapitalized, or has a spotty record of performance, an insurance policy may be the only safeguard afforded the buyer. In this time of expanding litigation, no firm can risk being in business without adequate insurance.

Other Documents

In a contradiction of terms, a customer on occasion may request an FDA guaranty letter for a technical grade product. Even though the buyer may have determined that the technical grade product is suitable for use in a food process, he cannot expect to receive a guaranty to this effect. As a matter of fact, any technical grade product could appropriately be labeled as ''not guaranteed as suitable for food use,'' leaving the decision entirely up to the buyer to determine the appropriateness of its use. As an added caveat, a supplier of technical grade product should avoid accepting a customer's specifications that in effect are equivalent to those given for a food product, thereby creating an implied guaranty of suitability for food use of the technical grade material.

Various legitimate demands made by customers often can be satisfactorily handled by the issuance of FDA guaranty letters. On rather questionable grounds, customers sometimes request Certificates of Analysis (COA) for each batch or lot of product received.[6] While such requests invariably can be met, they do contribute needlessly to the paper shuffle and entail shipping delays particularly from public warehouses. Unless discrepancies in product testing are suspected or are known to exist, an FDA guaranty letter should suffice in place of supplying COA's.

FDA guaranties can also be sent to satisfy request for ''*Salmonella* certificates'' or other guaranties such as freedom of contamination by polychlorinated biphenyls (PCB's). Inasmuch as FDA regards food containing *Salmonella* in any amount as adulterated,[7] an FDA guaranty is adequate evidence that the food manufacturer certifies such products covered by the guaranty as *Salmonella* negative. In the above situations and others like them, the food manufacturer may find the FDA guaranty letter to be a useful instrument for coping with a variety of concerns by his customers.

REFERENCES

1. "USDA Rule on Meat & Poultry Packaging," *Food Processing*, May, 1984, p. 14.
2. "Sulfite Label 'Continuing Guarantee' Program to Be Adopted by FSIS," *Food Chemical News*, December 1, 1986, p. 49.
3. "USDA to Require Guarantee Letters under Revised Sulfite Policy," *Food Chemical News*, January 19, 1987, pp. 44-45.
4. *Manual of Meat Inspection Procedures*, U.S. Department of Agriculture, Section 302.3, July 22, 1972 (change 79).
5. Lee Smith, " 'Equal Opportunity' Rules Are Getting Tougher," *Fortune*, June 19, 1978, pp. 152-156.
6. Anita L. Nolan, "Specifying Ingredients," *Food Engineering*, May, 1984, pp. 112-120.
7. "Microbiological Regulation," *Food Technology*, May, 1976, p. 42.

CHAPTER NINE
PRODUCT AND SAFETY DATA SHEETS

The transmittal of product information to customers must be done through formalized procedures in order to avoid regrettable consequences. Product data sheets, safety information, compositional data, and analytical reports are the equivalent of legal documents for which the seller can be held accountable. Therefore these data sheets and reports need to be prepared and reviewed with the utmost care. Mistakes in information supplied to customers can lead to the return of material and to costly settlements, not to mention a great deal of confusion and ill will.

PRODUCT DATA SHEETS

Data sheets can be considered extensions of product labeling and therefore are subject to the restrictions of the Federal Food, Drug, and Cosmetic Act (FD&C Act). Section 201(m) of the statute defines labeling not only as all written, printed, and graphic matter affixed to an article or its container but also all such information "accompanying" the article. Inasmuch as common practice is to enclose or send a data sheet with each sample or initial shipment of a product, such information would seem to fall within the scope of labeling. Moreover, since food advertising comes under the jurisdiction of the Federal Trade Commission (FTC), any promotional material like product data sheets would be affected by regulations adopted by this agency.

Trade Name

An example of a product data sheet for a leavening acid with the fictitious name Shurize is shown in Figure 9.1. It illustrates the principal features of such product literature. A trade name that does not infringe on the rights of any other manufacturer can be selected to refer to the product.[1,2,3,4] This identification is useful to the manufacturer and customer alike. The supplier is able to protect

CHEMICAL NAME	Monocalcium phosphate monohydrate
TRADEMARK	Shurize™
FORMULA	$Ca(H_2PO_4)_2 \cdot H_2O$
GRADE	Food Grade (Food Chemicals Codex)
DESCRIPTION	Monocalcium phosphate monohydrate is a white, free-flowing granular powder which is sparingly soluble in water.
SPECIFICATIONS	Calcium oxide: 22.2% min., 24.7% max.
	Loss on drying: 1.0% max.
	Arsenic (As): 3 ppm max.
	Fluoride (F): 25 ppm max.
	Heavy metals (as Pb): 30 ppm max.
	Lead (Pb): 5 ppm max.
TYPICAL PROPERTIES	Neutralizing value: 83
	pH (1% solution): 3.7
	Sieving: on 100 mesh 0.2%
	on 120 mesh 0.8%
	on 200 mesh 19.0%
	through 200 mesh 80.0%
	Bulk density: 65 lb. per cu. ft.
CONTAINERS	100 lb. net wt. multiwall bags
	Bulk shipment in rail cars
USES	Food leavening acid, dough conditioner, buffer, and dietary supplement
MANUFACTURER	Company name
	Post office address

10/1/87

FIGURE 9.1. SAMPLE PRODUCT DATA SHEET

hard-won markets while the consumer gains a ready handle for specifying the desired product. Any claim to a new trademark should be indicated by "TM" printed immediately following the brand name. This designation also indicates an intent to register the mark. Once registration has been received from the U.S. Patent and Trademark Office, the "TM" should be replaced by "R" within a circle. A trademark should be protected in this manner wherever it is used: on data sheets, labels, or in advertising.

Specifications

Product specifications must comply with any grade designation, e.g., Food Chemicals Codex, and in addition they should be consistent with such manufacturing specifications as adopted by the company. Considerable care needs to be taken in stating specifications. Values should be rounded to the nearest significant figures by accepted methods.[5] A product with an assay of 98.6 percent would meet a minimum specification of 99 percent but not 99.0 percent. Also, attention needs to be paid to the basis on which a result is reported. A dairy ingredient with 48 percent protein and 4.5 percent moisture would comply with a specification of 50 percent min. protein (dry basis), 5 percent max. moisture but not with the specification of 50 percent min. protein (as is) and 5 percent max. moisture. Bacteriological data are properly reported as "negative" rather than "absent" when laboratory methods of limited sensitivity do not show the presence of a given organism. These and other such refinements will avoid potential pitfalls or misunderstandings.

Typical Properties

A description of typical properties, defined as approximations to true values, can be very helpful in supplying additional information. For example, an amino acid profile for a vegetable protein product would indicate the quality of the available protein. Data on sieve analysis may have a direct bearing on caking tendencies or product reactivity. Bulk density is required to design packaging containers and storage facilities. The chief problem in providing typical properties in a data sheet is that they are sometimes misconstrued as legally binding specifications. On the other hand, just because a supplier cannot be held to the exact numbers supplied as typical values does not relieve him of the obligation to ship product that is representative of these data. A customer has a legitimate gripe when gross deviations occur.

Applications

A description of uses for a product draws attention to its valuable attributes and potential applications. The primary interest in using any functional food ingredient is the expected performance in a product. The chemical and physical properties of the ingredient are relatively less important. Thus, in selling an ingredient for a specific end use a supplier makes an implied warranty that it will function as indicated. To assist the customer in the proper use of a product, directions are often given. In this case any applicable warning statement about a potential hazard should be included or referenced.

Disclaimers

Notwithstanding a legal and moral commitment to supply a product that meets all specifications contained in a data sheet, suppliers like to include disclaimers, the wording of which appears to absolve them of any responsibility. Colloquially referred to as "boilerplate," such a disclaimer is usually placed in fine print at the end of the data sheet. Attorneys insist that some protection is provided by such statements. Particularly if the customer misuses a product, his grounds for recovering damages are diminished.

Copyright Statement

When a supplier wants to protect any information given in a data sheet from being reproduced, he is advised to add a copyright statement. All that is required is the copyright symbol, a "c" within a circle, followed by the owner of the copyright, and the year in which the material was prepared.[6] A copyright notice cannot be placed on old material that has been made public. It may be used, however, on substantial revisions. Such a statement will protect the text as well as graphic design on data sheets, labels, and promotional material.

Responsibilities of Groups

To ensure the proper preparation of product data sheets, a routine procedure should be established. One corporate group, say marketing, should assume the task of coordinating the preparation of data sheets. All responsible departments need to approve and initial the final draft. Each data sheet more than 18 months old since its last revision, for example, should go through the same review process before being reprinted. Only approved data sheets should be given to customers. Where data are preliminary, sketchy, or subject to rapid change, as for example with development products, a tentative or interim specification sheet may be prepared. Such a provisional sheet should clearly be labeled as tentative. If these steps are faithfully followed, accurate data will be disseminated to customers as required.

CUSTOMER SPECIFICATIONS

Not infrequently customers submit product specifications that do not conform with those issued by the supplier. The reason for such a request may be that the customer has a genuine need for a product that differs from that currently being supplied. More often the case is reduced to a situation where the customer lacks complete product specifications, has outdated figures, or simply is ignorant of standard practices. Regardless of the cause, every request for approval of customer specifications should be carefully considered and acted upon promptly.

Handling of Inquiries

The following procedure will assist management in handling all customer inquiries for product specifications. The Quality Assurance Department (QA) should be notified immediately of any such request. A copy of the written specifications should be sent to the plant which manufactures the product. The plant manager and the quality control supervisor acting jointly should recommend one of the following responses: 1. accept the specifications outright, 2. reject them with supporting reasons, or 3. modify them to meet certain objections. These recommendations are reviewed by QA, which makes the final decision on what action to take. Working through Marketing, QA notifies the customer of its position. If the customer comes back requesting amplification or further changes, the entire procedure can be repeated. All correspondence should meticulously record the customer's specification number, the date of its preparation or last revision, and a complete description of the product. A permanent file of all exchanges of memos should be maintained.

In answering a customer's request for approval of product specifications, certain guidelines are helpful. Whenever possible the submitted specification should be modified in order to conform with existing product specifications. If the number of different customer specifications grows unchecked, they will result in a set of plant manufacturing specifications which can be compared in complexity to a tree with many branches. Even the aid of a computer cannot reduce the multiplying chances for error, the escalating costs for special handling, and the lengthening delays in filling orders.

Special Requests from Customers

In reviewing customer specifications, care should be exercised to make sure that they are complete. Not only should product attributes and their limits be specified, but when in doubt the following criteria should be fixed: sampling techniques, test methods, container specifications, labeling, special storage conditions, shipping instructions, and protocol. The latter record outlines the agreed upon procedure for the submission of test results and other data. Often there is only one slight difference between customer "specs" and existing specifications. This nonconformity may appear to be minor at the time but can turn out to be quite nettlesome and the cause of many complaints. One common specification of this ilk is the request by customers to label all containers with their own unique product identification (I.D.) number. When the customer's clout is considerable, the supplier may feel compelled to comply, sometimes by printing I.D. numbers on all containers. Obviously an industry-wide numbering system would be welcome in this regard.

Customers often submit as a condition for approval of a supply contract, the right to inspect a manufacturer's facilities. Such requests are made with some justification, particularly in the case of new suppliers. On the other hand, the manufacturer may be reluctant to open his plant to inspection by outsiders. Under the circumstances, a compromise position might be for the manufacturer to withhold blanket permission to inspect his facilities but condescend to accept visitors by ''special permission only.''

Additional considerations should be kept in mind when reviewing customer specifications. All criteria must meet federal and state regulations. When, for example, a Food Chemicals Codex monograph exists for a given product it must be followed. A customer's specification and all data supplied by him should be treated as confidential. Written permission should be received before disclosing any proprietary information. A certain amount of give-and-take is necessary to resolve customer requests. More often than not, when a customer understands the reasons for changing or modifying his proposed specifications, he will accede to these suggestions.

SPECIAL RELEASE FOR SHIPMENT

No matter what the capability of a plant might be for producing quality product, minor processing upsets are inevitable. These variances will result in the manufacture of nonspecification product. A degree of flexibility is required for disposing of product that, save for a small defect, is perfectly good. Unless a procedure has been considered and adopted to release such off-standard material, judgmental errors may cause embarrassment or worse. Therefore a manufacturer should develop a special release procedure which delineates those exceptional circumstances under which nonspecification product may be shipped.

Nonspecification product, for the purpose of this discussion, is defined as material which deviates in a minor way from published specifications in data sheets, fails to meet a customer's purchasing specifications, or varies in a material respect from what the customer is used to receiving. The product in question, however, conforms to all federal and state regulations, and it is safe and wholesome. Generally, nonspecification product is so classified because it fails to comply with certain physical attributes such as color or particle size, or it deviates with respect to a performance standard like reaction rate. In most instances adjustments can readily be made by the customer to compensate for such deviations provided he is aware of the irregularity.

A special release for shipment can be handled by a simple but hard and fast procedure. A workable plan has been found to consist of the following elements.

1. A plant's quality control supervision may propose to release nonspecification product for shipment to a given customer. In a memorandum or on a report form he must explain the attribute which is out of specification and provide the test results for that attribute. He shall also indicate all relevant information such as the lot number, the customer's order number, and so forth.

2. The marketing manager, upon receipt of the proposal to release product, shall check with his customer, explaining the circumstances. Once permission is granted by the customer, the marketing manager shall confirm this approval in writing to the plant. Otherwise, if an OK is not received from the customer, the proposed release is rejected.

3. Product that is regraded, e.g., to "feed" or "technical" material, and which meets all of the requirements of the new designation, need not be reviewed under the procedure.

DISCLOSURE OF COMPOSITIONAL INFORMATION

A customer may require more information about a product than is normally disclosed by the manufacturer. Under these circumstances the information is usually conveyed to the customer under a nondisclosure agreement which requires all proprietary data to be held in confidence. Furthermore, the information may be used only for the intended purpose which is understood and agreed to by the manufacturer. While the legal procedure is effective in safeguarding trade secrets, it is cumbersome and too slow for routine or repetitive disclosures.

One situation that requires the frequent disclosure of proprietary information is the need to know product composition for the sake of labeling.[7] When a blended mix is used by a customer as an ingredient in one of his food products, he must know the percentage composition of the mix in order to list the ingredients of his own product in descending order of predominance on the label. In all but a few cases, approximate percentages for the mix composition will suffice. Therefore manufacturers commonly disclose compositional information in terms of ranges. A hypothetical blend would be reported as component X being between 0 and 10 percent, Y between 60 and 70 percent, and Z between 30 and 40 percent.

The disclosure of compositional data by ranges gives better precision if smaller ranges are used for trace components and larger ranges for macro-constituents. Thus, the following ranges have been proposed: 0-1, 1-5, 5-15, 15-30, 30-60, 60-100 percent. Generally, data given in these ranges will satisfy the needs of customers for labeling purposes. Collective ingredients such as flavors do not require the separate listing of each compound, and therefore compositional information is not required for these products. In other cases, as with intermediate mixes, the regulations may specify the disclosure of exact compositions so that these guidelines for reporting data in ranges would not apply. The disclosure of compositional information in terms of ranges is a workable compromise between the execution of nondisclosure agreements and the informal divulgence of formulations on an ad hoc basis.

REPORTS OF ANALYSIS TO CUSTOMERS

Customer service sometimes calls for the generation and submission of analytical reports. Certain accounts may insist on receiving Certificates of Analysis (COA) according to protocols for each lot of product shipped. Other segments of the food industry expect a degree of technical service that includes routine analytical support. One such business is devoted to the enrichment of flour where custom analyses are performed on raw materials in order to determine the correct level of vitamin fortification.

Analytical data are generally regarded as proprietary information. (cf. Chapter Six) To protect the confidentiality of these data a notice must be given with the submission of each report. Such a warning protects the manufacturer from the disclosure of any data to third parties without prior consent. Furthermore, while data are offered in good faith, most manufacturers are unwilling to provide a guaranty as to the accuracy or sufficiency of the information. The above understanding can be incorporated into a disclaimer worded as follows:

> We believe all information given in this report is accurate. It is offered in good faith but without guaranty. The supplier assumes no responsibility for the use of this information. Reports are submitted on a confidential basis. No reference to the report, the results or the supplier in any form whatsoever including but not limited to advertising, publication of statements, conclusions or extracts thereof may be made without the prior written approval of the supplier.

MATERIAL SAFETY DATA SHEETS

Material Safety Data Sheets (MSDS) complement product data sheets by providing information on product stability, storage conditions, disposal of unused material, spill handling, industrial hygiene, and general precautions for safe use. Data on toxicology are provided from the perspective of occupational safety and health. Even though food ingredients have the imprimatur of FDA, few of these materials are completely innocent under industrial conditions of use. Strong acids and bases are highly corrosive and may cause severe burns. Powdered combustible products such as cocoa, cornstarch, dry milk, and sugar may lead to explosions if their dusts are ignited by sparks or other means.[8] With poor ventilation, tricalcium phosphate by nature of its low solubility can result in the irritation of the upper respiratory tract, possibly causing nosebleeds. Other ingredients, for instance ammonium bicarbonate, can give off noxious gases in warehouse fires or other blazes.

OSHA-20 Form

For years many manufacturers have supplied their customers with an MSDS for each of their hazardous products. This service was provided as a measure of good will since manufacturers, until a short while ago, were not under any legal obligation to provide safety information. The full responsibility for worker safety was placed squarely on the employer who had to obtain the necessary data as best he could. Lacking an industry approved format, manufacturers who chose to provide safety data frequently resorted to the use of the OSHA-20 form. This form was available from the U.S. Department of Labor (USDL). Even though it was designed exclusively to comply with regulations for ship repairing, the OSHA-20 form covered the points of principal concern. The required information is given in Section 6(b)(7) of the Occupational Safety and Health Act, passed in 1970, which states:

> Any standard promulgated under this subsection shall prescribe the use of labels or other appropriate forms of warning as are necessary to insure that employees are apprised of all hazards to which they are exposed, relevant symptoms and appropriate emergency treatment, and proper conditions and precautions of safe use of exposure.

Right-to-Know Statutes

Until recently the Occupational Safety and Health Administration (OSHA) delayed action on the implementation of Section 6(b)(7). Before enforcement procedures were eventually taken by OSHA increasing numbers of state and municipal governments began passing right-to-know statutes. By 1982 nine states and the city of Philadelphia had such regulations on the books, and more legislative districts were seriously considering the passage of similar bills.[9] One of the most comprehensive laws was passed by California in 1980. Known as the Hazardous Substances Information and Training Act, this bill provided for many sweeping reforms that included not only worker safety but also environmental protection.[10] With the proliferation of state and municipal laws, none of which were the same, manufacturers began to realize that one national, all-encompassing standard, however stiff it might be, would be preferable to interminable local legislation.

Contents of an MSDS

On November 25, 1983, OSHA promulgated final regulations for hazard communication (29CFR1910.1200). The regulations provided that, instead of relying on labels to pass along safety information, MSDS's were to be the primary means for transmitting this data. The use of a label would be limited to identifying the name of any hazardous chemical, supplying an appropriate hazard warning, and giving the name and address of the manufacturer. The extensive and necessarily complex material provided in an MSDA would be explained to employees through formal training programs. Under the new regulations manufacturers are obligated to send an MSDS with each initial shipment of a product. This same provision applies to distributors, who must also supply an MSDA, obtained from the manufacturer, to each of their customers. In order to ensure that a required MSDA is available at every workplace it is incumbent on the employer who purchases a material to request any missing data sheets.[11]

The contents of an MSDA are specified in the regulations as follows:

1. The identity used on the label . . .;
2. Physical and chemical characteristics . . .;
3. The physical hazards of the hazardous chemical, including the potential for fire, explosion, and reactivity;
4. The health hazards . . .;
5. The primary route(s) of entry;
6. The OSHA permissible exposure limit, ACGIH [American Conference of Governmental Industrial Hygienists] Threshold Limit Value and any other exposure limit . . .;
7. Whether the hazardous chemical is listed in the National Toxicology Program (NTP) *Annual Report on Carcinogens* . . .;
8. Any generally applicable precautions for safe handling and use . . .;
9. Any generally applicable control measures . . .;
10. Emergency and first aid procedures;
11. The date of preparation of the material safety data sheet or the last change to it; and
12. The name, address and telephone number of the chemical manufacturer, importer, employer or other responsible party preparing or distributing the material safety data sheet. . . .

Supplemental Information

Some manufacturers consider the above OSHA standards as minimum requirements for the content of an MSDS. They make use of the MSDS as a vehicle for including additional information that may be helpful to customers. An illustration of an MSDS is shown in Figure 9.2. The most difficult section to prepare is the one on toxicology, which still is very much an arcane science.[12,13,14] In assessing toxicological effects, reference should be made to reports of adverse reactions by employees. A file of these records must be kept for 30 years as provided by the Toxic Substances Control Act.[15] This type of information is being supplemented through a mounting effort by industry and government to expand toxicological research.[16,17]

This product safety information is principally directed toward managerial and professional personnel. It is intended only as a starting point for the development of safety and health procedures including employee educational and training programs.

I.   NOMENCLATURE

Synonyms: monocalcium phosphate monohydrate;

                   calcium phosphate monobasic monohydrate

Trademark: Shurize™

Chemical Abstracts Service (CAS) Number: 10031-30-8

II.   PHYSICAL AND CHEMICAL PROPERTIES

Grade: Food Grade

Formula: $Ca(H_2PO_4)_2 \cdot H_2O$

Physical State: White granular powder

Decomposition point: Loses water of hydration at 70ºC (158ºF).

                    Loses water of composition at 200º C(392ºF).

Solubility: 1.8g per 100g water at 30ºC

pH: 3.7 (1% aqueous solution)

Purity: meets Food Chemicals Codex specifications

III.   CHEMICAL REACTIVITY

The product is relatively inert. It forms an acidic solution in water.

IV.   STABILITY

The product has a shelf life of over one year when properly packaged and stored.

V.   FIRE HAZARD

This product is not combustible, nor will it support combustion. Toxic gases are not liberated by the material when heated by fires.

VI.   FIREFIGHTING MEASURES

Use standard firefighting techniques in extinguishing fires in the presence of this product. Water, dry chemical, foam, carbon dioxide or other suitable suffocation agents may be employed.

As in the case of any fire, prevent human exposure to fire, smoke, or products of combustion. Wear full-face, self-contained breathing apparatus and impervious clothing, such as gloves, hoods, suits, and rubber boots. Evacuate all nonessential personnel from the impacted area.

FIGURE 9.2. SAMPLE MATERIAL SAFETY DATA SHEET

(*continued*)

VII.   TOXICOLOGY

Monocalcium phosphate is listed by the Food and Drug Administration as "Generally Recognized as Safe" (GRAS) for use as a food additive. (21CFR182.1217) Ingestion, however, of large quantities may cause abdominal discomfort. Concentrated aqueous solutions are strongly acidic; contact with the eyes, mucous membranes or repeated skin contact may cause local irritation.

Ingestion

Ingestion of large quantities may produce non-specific irritation of the mouth, throat, and gastrointestinal tract; nausea, vomiting, cramps, and diarrhea.

Skin Contact

Local irritation of the skin may result from contact with the powder or concentrated aqueous solutions.

Eye Contact

Local irritation of the eyes may result from contact with powder, granules, or aqueous solutions.

Threshold Limit Value (TLV)

The American Conference of Governmental Industrial Hygienists (ACGIH) has not established a TLV for monocalcium phosphate.

VIII   FIRST AID

If a known potentially hazardous exposure occurs or is suspected, immediately initiate the recommended procedures below. Contact a physician, a neighboring hospital, or the nearest Poison Control Center to obtain additional medical advice. For further information call collect day or night, CHEMTREC at (800) 424-9300.

Ingestion

If swallowed in large quantities immediately dilute the ingested material by giving large quantities of water. Induce vomiting by gagging the victim with a blunt object placed on the victim's tongue. Continue fluid administration until vomitus is clear. Never give anything by mouth to an unconscious person. Call a physician or the nearest Poison Control Center immediately.

Skin Contact

Wash all affected areas with plenty of soap and water. Remove contaminated clothing. Seek medical attention if irritation occurs.

Eye Contact

Flush the eyes with running water for 15 minutes. Hold the eyelids apart during the rinsing to ensure flushing of the entire surface of the eye and lids. Obtain medical attention if eye irritation occurs.

FIGURE 9.2. SAMPLE MATERIAL SAFETY DATA SHEET (*continued*)

Inhalation

Remove from contaminated atmosphere. Seek medical attention if respiratory irritation occurs.

IX.   INDUSTRIAL HYGIENE

Ingestion

All food should be kept in a separate area, away from the working location. Eating, drinking and smoking should be prohibited in areas where there is a potential for significant exposure to this material. Before eating, thoroughly wash face and hands.

Skin Contact

Skin contact should be minimized through the use of gloves and suitable long-sleeved clothing.

Eye Contact

Eye contact should be prevented through the use of chemical safety glasses, goggles, or a face shield depending on exposure potential.

Inhalation

If use conditions generate dust, the material should be handled in an area with local exhaust ventilation or in the open (e.g. outdoors). Where adequate ventilation is not available NIOSH-approved respirators of adequate design should be employed.

X.   SPILL HANDLING

No emergency will result from minor spills, but spilled material should be cleaned up promptly. Make sure all personnel involved in the cleanup follow good industrial hygiene practices (refer to Section IX). Sweep up or vacuum spilled material. **Do not return spilled product to its original container.** Contaminated material should be discarded in a manner that will not adversely affect the environment. Flush the spill area with water to remove any residue.

XI.   MATERIALS OF CONSTRUCTION

Because glass may shatter under impact, it should be avoided as a material of construction. Type 316 stainless steel is corrosion resistant and therefore is satisfactory for this chemical and its solutions. Metal traps, sieves, and detectors should be used at critical control points of a process using this material.

XII.   STORAGE REQUIREMENTS

This material is a food ingredient intended for human consumption. Therefore it must be held under sanitary conditions and kept isolated

FIGURE 9.2. SAMPLE MATERIAL SAFETY DATA SHEET (*continued*)

from all toxic and harmful substances. The product should be stored
in a cool, dry place. Keep containers closed when not in use. Exercise
due caution to prevent damage to or leakage from containers.

XIII.   DISPOSAL OF UNUSED MATERIAL AND CONTAINERS
Residual quantities of unused material do not require special handling
when disposing of empty containers. Such containers may be discarded
with the general trash or incinerated in approved facilities. Larger
quantities of the material that cannot be used or reprocessed should be
discarded, after checking with local, state, and federal authorities, in a
manner that will not impact on the environment.

XIV.   MANUFACTURER'S NAME AND ADDRESS
Company name
Post-office address
Telephone number
Date of MSDS preparation or latest revision

All information contained herein is furnished without
warranty of any kind. Employers should use this
information only as a supplement to other
information gathered by them. They must make their
own independent determinations of the suitability and
completeness of data from all sources to assure the
proper use of this material as well as the safety and
health of their employees.

FIGURE 9.2. SAMPLE MATERIAL SAFETY DATA SHEET (*continued*)

Under the first aid section of an MSDS the hot line of the Chemical Transportation Emergency Center (CHEMTREC) is given. This service, which was inaugurated in September, 1971, has proven to be vital in responding quickly to emergencies.[18,19] With the installation of a new computer program in 1986, CHEMTREC is now able to provide printouts of vital data over telephone lines. Information is stored on more than 1,700 chemicals and eventually will be expanded to cover over 4,000 compounds.[20] Another section of the MSDS relates to industrial hygiene which is not without controversy. There is an ongoing debate concerning the trade-off between installing more pollution abatement equipment versus the wider use of personal protection devices. Such worker safety gear as respirators and gloves that are used must be certified by the National Institute of Occupational Safety and Health (NIOSH).[21,22]

Assistance can be obtained from several sources in preparing an MSDS. The Chemical Manufacturers Association has issued a guide to the preparation of MSDS's.[23] By 1978 this industry organization had published some 57 safety monographs covering the more hazardous substances.[24] An international effort to disseminate safety information is underway through the auspices of the International Register of Potentially Toxic Chemicals (IRPTC) in Geneva, Switzerland.[25] One of the primary concerns in the preparation of MSDS's is providing quick recognition of potential hazards. To this end, suggestions have been made to adopt a set of symbols to alert workers.[26,27] Although the MSDS format is still evolving, its basic requirements have been established.

REFERENCES

1. "Trademarks under Fire," *Dun's Review*, September, 1978, pp. 30, 36, 104-106.
2. Stephen Solomon, "Formica's Fight for its Own Name," *Fortune*, September 10, 1979, pp. 134-138.
3. "Trademarks: the Other Mess," *Chemical Week*, February 18, 1981, pp. 44-46.
4. "Trademarks Why Companies Guard Them More Tightly," *Chemical Week*, January 13, 1982, pp. 30-33.
5. *Food Chemicals Codex*, 2nd ed., National Academy of Sciences, Washington, D.C., 1972, p. 4.
6. Charles R. Goerth, "The Legal Impact," *Packaging Digest*, December, 1981, pp. 28-30.
7. John F. Wintermantel, "Working Relationships between Users and Suppliers of Raw Materials," *Food Technology*, April, 1984, pp. 113-114.
8. Sid Clark, "Preventing Dust Explosions," *Chemical Engineering*, October 5, 1981, pp. 153-154.
9. *Federal Register*, March 19, 1982, pp. 12092-12124.
10. "California OSHA Lists Hazardous Chemicals," *Chemical & Engineering News*, February 8, 1982, p. 8.
11. *Federal Register*, November 25, 1983, pp. 53340-53346.
12. J. C. Kirschman, "Safety Evaluation Is a Risky Business," *Food Product Development*, July-August, 1977, pp. 56-58.
13. Charles J. Krister, "Safety Evaluation . . . It Ain't All Science," *Chemtech*, November, 1979, pp. 668-672.
14. Kenneth J. McNaughton, "The ABCs of Occupational Skin Disease - Part I," *Chemical Engineering*, March 22, 1982, pp. 147-150.
15. "Toxic Substances, Toxicology, and the Chemical Engineer," *Chemical Engineering*, April 24, 1978, pp. 70-89.
16. William Reddig, "Industry's Preemptive Strike Against Cancer," *Fortune*, February 13, 1978, pp. 116-119.
17. Raul Remirez, "Industry Toxicology Group Changes Leaders," *Chemical Engineering*, February 23, 1981, pp. 34-36.

18. "Hazardous Materials Emergency Systems Linked," *Chemical & Engineering News*, March 24, 1980, pp. 7, 8.
19. "Emergencies: Chemtrec Isn't Enough," *Chemical Week*, July 7, 1982, p. 23.
20. "CHEMTREC has a HIT," *Chemical Week*, September 17, 1986, pp. 18-19.
21. "NIOSH Challenges a Respirator's Efficacy," *Chemical Week*, July 1, 1981, pp. 12, 13.
22. "NIOSH Finds the Gloves of Workers Inadequate to Handle Toxic Substances," *Chemical Marketing Reporter*, July 6, 1981, pp. 5, 23.
23. *A Guide for the Preparation of Chemical Safety Data Sheets*, Chemical Manufacturers Association, Washington, D.C., 1978.
24. *Publications List*, Chemical Manufacturers Association, Washington, D.C., 1978.
25. Dermot A. O'Sullivan, "Global Toxic Chemicals Register Takes Shape," *Chemical & Engineering News*, May 10, 1982, pp. 25-27.
26. "ICI Develops Graphic Approach to Hazard Communication," *Chemecology/ Special Report*, May, 1981, p. 8.
27. "Pinning a Label on Labeling," *Chemical Business*, January 11, 1982, p. 6.

CHAPTER TEN
COMPLAINT HANDLING

The frequency of customer complaints is the single, most reliable indicator of a breakdown in quality control. These reports are the surest signs of failures in production or distribution. Over a period of time the level of complaints provides a running tally of the progress being made in correcting problems and improving operations. Because complaints are generated externally to an organization, they are the most objective appraisal of performance. These data are also the most relevant since they are a barometer of customer satisfaction. For these reasons, a food manufacturer needs to adopt a formalized procedure for handling complaints, and it must maintain an accurate record of all complaints received.

REPORTING

A formalized procedure for handling complaints is indicated by the flow diagram in Figure 10.1. The first knowledge of a complaint comes when a customer notifies a company representative of a problem. Only certain individuals are qualified to receive the particulars of a complaint. Such a person, called an Initiator for the purpose of this procedure, may be a sales representative, product manager, regional manager, national accounts manager, marketing manager, quality assurance manager, quality control supervisor, technical sales representative, or manufacturing coordinator. A number of employees, however, are not authorized to accept the disclosure of a complaint. These personnel, including secretaries, telephone operators, trainees, administrative assistants, and technicians, should obtain just enough information, i.e., the name of the caller, his firm, location, telephone number, and product involved, in order to refer the message to the proper individual within the company. The receiver of the call should not tell the customer to contact someone else. Shifting responsibility to the customer is not only discourteous, but any delay in the follow-up could lose precious hours.

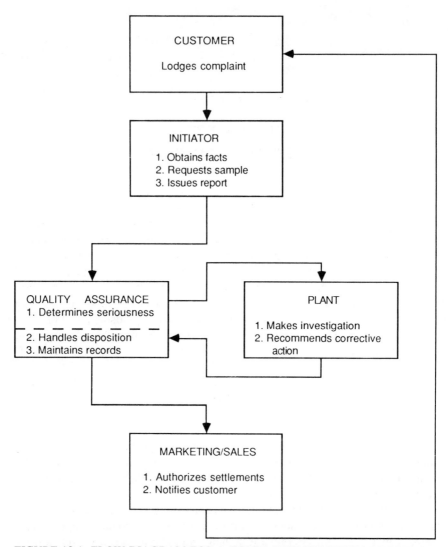

FIGURE 10.1. FLOW DIAGRAM FOR A COMPLAINT HANDLING PROCEDURE

Sometimes a complaint may be received through other channels. Particularly in the case of a consumer complaint, a government agency may be the first party to hear about an incident. In a 1984 policy directive, the U.S. Department of Agriculture (USDA) promised to forward to the manufacturer all such information received from consumers of meat and poultry products.[1] Alternatively, the Initiator may get a written or oral communication from an attorney or physician claiming to represent a client. Under these circumstances the complaint should be presumed to involve product liability with all the ramifications that this situation entails.

Initiator's Role

When the Initiator makes contact with the customer, he must keep in mind a list of do's and don't's. The principal assignment of the Initiator is to obtain as much factual information surrounding the complaint as possible, while the knowledge of the complaint is fresh in the mind of the customer and before positions have hardened. The initial contact is the best time to gather helpful specifics which later on may be overlooked or suppressed. Certain information is an absolute must. The requisite data include a complete description of the product, the code date (lot number), and preferably the invoice number or, if this item is missing, the date of receipt of the shipment. Other facts can be very useful. If a quality defect is reported, numerical results of laboratory tests should be obtained. The exact amount of the unacceptable product should be recorded. Finally, if the customer has expended any extra work or additional expense, the number of man-hours or the out-of-pocket cost needs to be pinpointed at this early stage.

Regardless of the circumstances relating to the complaint, the Initiator should scrupulously pay attention to the following warnings:

- Do not admit fault.
- Do not promise any settlement or performance.
- Do not promise to furnish analytical results on samples received.
- Do not mention any insurance coverage.

Besides jeopardizing the interests of the manufacturer, any admissions or promises during this opening contact can only confuse the issue. The Initiator must defer any question of equitable treatment to his management for resolution.

Nature of the Complaint

At the outset, the Initiator should attempt to verify the true nature of the complaint. First, he should determine whether the party filing a complaint is indeed a customer. Anyone who has purchased product either directly from the manufacturer or through a bona fide distributor would be considered a customer. If an intermediary has repackaged, treated, or otherwise processed the product, then the manufacturer is justified in disclaiming any responsibility. In this case, such assistance that may be offered by the manufacturer is not tied to the complaint.

Other nuances in the opening communication with the customer need to be determined. Is the customer in reality lodging a complaint or is he merely looking for guidance and help? He may only be inquiring about certain properties of the product or seeking instructions on how to use it. A complaint, however, implies that the manufacturer somehow failed to meet the customer's expectations. It is usually accompanied by a request for action on the part of the manufacturer. The customer may request the replacement of part or all of a shipment, he may seek approval to return product, or he may ask for a credit memo. In those instances where the customer is only advising or warning the manufacturer of a problem, he expects, at the very least, an investigation by and a report back from the manufacturer.

In order that a correct response will be forthcoming, the manufacturer needs to classify complaints by type. Certain complaints on billing, pricing, and scheduling fall outside the domain of the quality assurance function and are best handled by interoffice memorandum. A quality complaint can be defined as any one involving the properties of the product, the integrity of the package, or the accuracy of the labeling. For the sake of handling, these three subcategories of complaints, namely, product, packaging and labeling, may be considered separately. The plant, however, must answer for all of them. Any complaint concerned with the condition of a product as a result of mishandling by a common carrier or a public warehouse is considered to be a service complaint. Finally, a liability complaint is one which involves any loss due to bodily injury or property damage. The exact nature of a complaint may not become entirely clear until an investigation is well underway, but the Initiator should make some attempt at defining it in his report.

Filing a Report

Within 24 hours of receiving a customer complaint, the Initiator should report the facts by telephone to the Director of the Quality Assurance Department (QA). Speed is especially critical when a government agency is already involved, when the possibility exists that a product recall may be necessary, or if the complaint relates to product liability. By taking quick action the manufacturer often can limit its losses and at the same time protect the interests of its customers. Complaints of a less serious vein do not call for such a rapid response; however, the decision as to what constitutes a serious complaint should be made by QA, not by field personnel. When the Initiator delays notifying QA, he in essence is arrogating managerial responsibility.

After verbal communication with QA, the Initiator should prepare a detailed written account of the complaint. As a rule of thumb, this report should be distributed within two working days from the time of receiving the complaint. For the purpose of reporting the information, a complaint form, similar in content to that shown in Figure 10.2, should be used. When filling out the form, the Initiator ought to distinguish the facts as he knows them from the assertions of the customer. If additional space is required or if explanatory material is available, these addendums may be attached to the form.

In order to monitor and coordinate the handling of complaints as well as to analyze the collective results, every complaint needs to be numbered. In larger firms there may be several regional offices or locations where complaints might be received. In this case the simplest system for numbering complaints is to assign a location code for each office. Using such a method, complaint No. K12-84 might stand for the twelfth complaint received by the Kansas City office in the year 1984. A log would be kept in this regional office of all complaints reported to

CUSTOMER COMPLAINT REPORT

Complaint No.

DISTRIBUTION:

Plant Manager_____   Director Marketing_____
Plant Laboratory_____   Director Quality Assurance_____
Director Manufacturing_____   Regional Sales Manager_____

Customer Name & Address	Plant	Ship Date
	Product	Complaint Date
	Code Date	Invoice No.
Distributor Name & Address	Quantity	Carrier
	Has a sample been obtained? Yes ☐ No ☐	

DESCRIPTION: Quality ☐   Service ☐   Liability ☐

By_____   Date_____

INVESTIGATION:

CORRECTIVE ACTION:

By_____   Date_____

DISPOSITION:

Allowed   ☐
Disallowed   ☐
Indeterminate   ☐   By_____   Date_____

FIGURE 10.2. CUSTOMER COMPLAINT REPORT

it. Missing reports can easily be tracked down by staff members at headquarters making use of this system of consecutively numbering complaints received at each location. Any complaint not recorded on a complaint form and given a number stands the chance of being ignored by personnel assigned the task of investigating complaints.

Copies of the completed complaint form are sent to everyone on the distribution list as well as to other individuals or departments that may have an interest. Thus, any liability complaint would be circulated to the Law Department. The Plant Manager and the Quality Control Supervisor receive their copies of the report so that they can immediately begin their investigations. The other people on the distribution list need to be kept apprised of progress at each stage in the handling of the complaint.

Sample Request

Whenever a complaint concerns the quality or performance of a product, a sample of the material in question should be requested from the customer. This sample should be sent to the plant control laboratory which is responsible for the particular product. It should be clearly labeled, and to avoid possible confusion, the laboratory should be put on the lookout for it. As hard as it is to believe, samples are too often contaminated by the method used in taking them or by the containers in which they are placed. The risk of contamination should be noted whenever there are signs of slipshod procedures. Occasionally, material has been attributed to the wrong supplier, as for example when a sample is removed from a storage bin. A sound policy for the manufacturer to follow is to dispatch a representative to the customer's facility to supervise the taking of any samples and to note the conditions of the place. If a sample is not available, for instance if the product has been completely used by the customer, this omission is prima-facie grounds for disallowing the complaint.

INVESTIGATION

Responsibility for handling a quality complaint involving either the product, packaging, or labeling rests with the Plant Manager or his designee. Generally, complaints about packaging or labeling will be referred to the head of the Shipping Department for investigation. Logically the Control Laboratory will look into any report of nonspecification product. All production records, control charts, employee time sheets, maintenance logs, laboratory notebooks, complaint files, and other relevant materials must be accessible to those persons conducting the investigation. Retained samples are invaluable for rerunning critical tests. Investigative techniques are not unlike those used by a forensic laboratory where clues are uncovered by painstaking efforts.

The length of an investigation and the effort put into it will be dictated by the complexity of the problem. Therefore, to assign an inflexible time limit for the completion of an investigation is unrealistic. Of course, those complaints which are most pressing should receive priority, even to the extent where results are reported on a day-by-day basis. Often the most difficult kind of problem to resolve is one related to the performance of a product. Application studies may be needed to uncover the difficulty. Since application equipment usually is not available in a plant control laboratory, complaints involving performance frequently must be referred to the company's research department or to an independent testing laboratory. Delays, therefore, in investigating performance complaints are not uncommon.

The results of an investigation are entered in the appropriate space on the Customer Complaint Report. Equivocations in this report are not appreciated.

The Plant Manager must indicate the causes of the problem. He must answer the question, What went wrong and why? If he does not know for certain what happened but has a notion, he should offer his comments, clearly stating that they are his interpretation of the facts. He should report as facts only those matters within his knowledge. To complete the report, the Plant Manager must outline what corrective action is planned, who is responsible for directing it, and when it will be completed. Obviously, if the plant finds in its investigation that the complaint was unfounded, no preventive action would be recommended.

Service Complaints

Service complaints will require a somewhat different approach for their investigation. Whenever a customer receives damaged goods which are suspected as having been mishandled by the common carrier, the following prescribed steps should be taken by him. He should:

1. Record the seal numbers and state whether the seals have been broken or are intact.
2. Not unload the car in question.
3. Request an immediate inspection by the carrier.
4. Take photographs of the damaged goods.
5. Salvage as much product as possible once the investigation by the carrier has been completed.
6. Promptly file a claim with the carrier, making sure to include the necessary supporting documents.[2]

Simultaneously with contacting the carrier, the customer should notify the manufacturer of any service difficulty. The originating plant can then pursue an investigation at its end. Evidence should be presented by the manufacturer that the empty car had been inspected and that the shipment was in good condition, properly loaded, and contained the specified amount of material. To protect itself, the manufacturer should take pictures on a regular basis of the contents of each outbound car. In addition, an impact indicator might be included to document any rough handling, as for example when railroad cars are shunted.[3] The Food and Drug Administration (FDA) has approved the design of an indicator which reveals the thawing of frozen products.[4]

Two of the major headaches in determining the culpability of service complaints are caused by transshipments and stop-off shipments. Less-than-truckload quantities frequently are reloaded onto new vans at terminals for transshipment. Stop-off shipments are made to intermediate consignees for delivery of partial loads. In either situation, product is at the mercy of the carrier or the intermediate consignee. Not only can containers be damaged, but product may be stored under insanitary conditions or part of the shipment may be lost or stolen.

Other service complaints can be traced to delinquencies committed by public warehouses. Depending on assigned responsibilities within the corporation, either the Distribution Department or the Regional Sales Manager may be given the task of investigating the complaint. A thorough inspection of the warehouse should be made in order to determine the cause of the problem. Only accredited food warehouses should be used for the distribution of food products.

Even though validated service complaints are not the direct responsibility of the manufacturer, they reflect poorly on its management. In order to minimize the frequency of these complaints the supplier is advised to keep a detailed tabulation of them. As warranted, those carriers and warehouses that are repeat violators should be blacklisted. Another effective step is to persuade more customers to order full truckload quantities, thus minimizing the potential hazards of transshipment. Where these efforts fail, customers may be encouraged to order product through local distributors.

DISPOSITION

An effective procedure for handling complaints includes a resolution or disposition step. After the completion of the investigation, QA reviews the report and determines whether the complaint is valid (justified), invalid, or indeterminate. As a legal nicety, the terms "allowed" and "disallowed" may be substituted for "valid" and "invalid" thereby avoiding the possible admission of fault. In this role, QA acts as a sort of referee in the seemingly never ending contention between Marketing and Production. Marketing is dedicated to better serving its customers while Production is committed to improving operating efficiencies. A degree of statesmanship is thus required by QA in deciding the merits of complaints.

The disposition of complaints is more than an exercise in scorekeeping. If a complaint is allowed, such a determination supports and even demands a program of corrective action by Production. An "allowed" disposition also is a signal to Marketing that restitution to the customer is justified. On the other hand, a "disallowed" complaint does not prevent Marketing from accepting returned material, giving credit, or replacing damaged or lost product. Such decisions are up to Marketing, which has to consider other factors, such as customer relations and competitive pressures. In fact, Marketing need not wait until the completion of the investigation to decide its own course of action, although it should be able to account for its stand. When a liability claim is settled, the manufacturer is cautioned to request a release (Figure 10.3) from his customer. Whatever the settlement, maximum effectiveness of this complaint procedure cannot be achieved if QA too frequently arrives at an "indeterminate" disposition.

Special Needs of Consumer Products

No two food manufacturers will have identical procedures for handling complaints. A producer of consumer products very likely has a Consumer Affairs Department. Other companies may provide for a Technical Sales Department or

In consideration of the payment of $_____   the

undersigned_____ (Buyer) hereby

releases and   forever discharges _____(Seller)

and   _____ (Distributor) from any

consequences   or claims now or hereafter known arising out of the

following circumstances:

_____

_____

_____

_____

Dated this _____   day of _____, 19__

By: _____

_____

(Signature and title of

representative for buyer)

FIGURE 10.3. RELEASE

a Sales Development Department. Such groups that are geared to take care of customer relations often are assigned certain responsibilities otherwise given to Quality Assurance and to Marketing/Sales for handling complaints. Where large numbers of consumer complaints are received, individual ones, except in rare instances, would not be investigated separately. Instead, they would be characterized by product and attribute, and would be lumped with similar complaints in order to provide feedback information to Quality Assurance and to the Production Department.

Responses to consumer complaints ordinarily are more perfunctory than replies to complaints received from industrial accounts. A consumer may be sent cash value coupons, a replacement product, or a form letter depending on the merit of his complaint and the manufacturer's policy. Some manufacturers facilitate complaint handling by supplying toll-free "800" telephone numbers on product labels and in promotional material. As a rule, a manufacturer should check each complaint to see if the same customer has complained before. When two or more complaints are received from the same person the chances are greater that the objection is invalid. In such cases, a special response is advisable. A consumer complaint that is difficult to resolve may be turned over to an outside organization, such as the National Food Processors Association, for handling in an equitable manner.[5]

ANALYSIS

Until the data on complaints have been collated and analyzed, their significance cannot be fully grasped. A form similar to the one shown in Figure 10.4

No.	Plant	Product	Customer	Description	Disp*	Date Recd	Date Disp

* Disposition: A = Allowed, D = Disallowed, I = Intdeterminate

FIGURE 10.4. CUSTOMER COMPLAINT RECORD

is useful for recording the important details, including the plant, product, customer, description, disposition, and the dates when the complaint was received and resolved. In cases where the number of complaints are substantial, this information can be handled by a digital computer. Software is available for personal computers to store, modify, retrieve, sort, print and otherwise manipulate the information.[6]

Organizing Complaints

In a computer program designed to handle a data bank, each complaint, which is identified by its number, constitutes a separate record. This record can be likened to a single folder in a filing cabinet. The individual items, such as the product name, customer identification, etc., contained in the folder or so-called record make up the different fields. The computer, when instructed to do so, can sort by fields, e.g., retrieve and print all complaints on a given product. It can also calculate the required times in handling complaints by subtracting the dates of receipt from the corresponding completion dates. Thus, by means of a computer a large amount of information can be rapidly processed and compactly stored.

Further insight into operations can be obtained by analyzing complaints according to the types of problems. By abstracting information from the descriptions of complaints, a report like the one shown in Figure 10.5 can be prepared. It arranges the complaints by products and also by attributes, namely, color, microbiology, granulation, purity, packaging, service, and so forth. These breakdowns very quickly indicate troublesome areas of operation. At the same time, they highlight managerial responsibilities for taking corrective measures.

	Acidulant A	Acidulant B	Alkali	Dough Conditioner	Emulsifier	Flavor Enhancer	Leavening Acid	Soy Protein	Total
Color									
Microbiological									
Sanitation									
Granulation									
Foreign Matter									
Purity									
Packaging									
Lableing									
Performance									
Total									
Service									
Delays									

FIGURE 10.5. ANALYSIS OF CUSTOMER COMPLAINTS

The data from a report as shown in Figure 10.5 can be used to prepare bar charts, which in this application are known as Pareto charts. These graphs are named after Vilfredo Pareto, an Italian economist, who in the late 1800's recognized certain disparities in statistical data. Specifically, he noted that 80 percent of the wealth was owned by 20 percent of the people, and that 80 percent or more of the crimes were committed by 20 percent or less of the population. Hence, the Pareto principle is often referred to as the 80/20 rule. As applied to quality control, this rule becomes quickly obvious: certain problems are found to cause a disproportionate number of complaints.[7]

Frequency of Complaints

Complaint data can be plotted to indicate the frequency of occurrence over a period of time. This information can be presented to show monthly, quarterly, or yearly results. Adjustments may be made to correct the data for significant changes in the sales volume of a product. Different parameters may be picked as being most relevant, for example, the total number of complaints, only the number of allowed complaints, or only quality complaints. The number of delays in handling complaints can be reckoned by counting all of those complaints which took longer to resolve than an allotted period of time. The information on delays is helpful in spotting tardiness in either reporting or investigating complaints.

When the frequency of customer complaints is graphed, a sharply fluctuating pattern is generally obtained as shown in Figure 10.6. These swings are caused by many random events such as vacations, mail deliveries, the weather, and sales promotions, all of which have nothing to do with quality. Therefore these factors need to be isolated or corrected in order to discern meaningful trends. By means of regression analysis, the slope of the correlated data can be determined and control limits can be placed around this slope.[8] Computer programs are now available to handle large quantities of statistical data on complaints in order to determine time trends and confidence limits. These delineations, when indicated on the frequency chart, provide a ready means to forecast performance.[9]

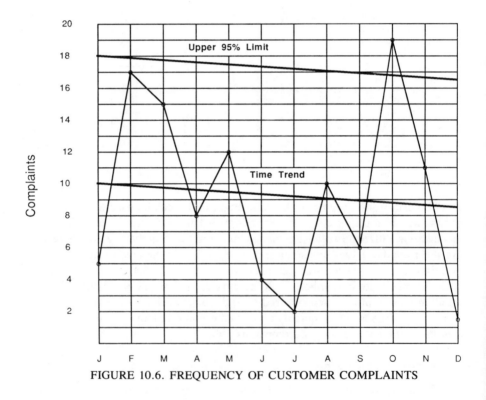

FIGURE 10.6. FREQUENCY OF CUSTOMER COMPLAINTS

COSTS

A study of the costs of handling and settling complaints reveals some fascinating relationships. Table 10.1 indicates the direct costs of three representative

TABLE 10.1
DIRECT COSTS OF REPRESENTATIVE CUSTOMER COMPLAINTS

I. DAMAGED CONTAINERS
Assumption: 1 percent of shipment was replaced.

Sales & Administration, 1 man-day	$ 300
Investigation, 1 man-day	300
Product, 400 lb. @ 50¢	200
Freight, 10¢ per lb.	40
TOTAL	$ 840

II. PRODUCT QUALITY
Assumption: Entire shipment was rejected: product had no salvage value.

Sales & Administration, 2 man-day	$ 600
Investigation, 4 man-day	1,200
Product, 40,000 lb. @ 50¢	20,000
Freight, 1¢ per lb. each way	800
TOTAL	$ 22,600

III. PRODUCT LIABILITY
Assumption: Product was used at 1 percent level in customer's cake mix; the mix had no salvage value

Sales & Administration, 10 man-day	$ 3,000
Investigation, 20 man-day	6,000
Claim, 4 million lb. @ 25¢	1,000,000
TOTAL	$1,009,000

types of customer complaints. The data were arbitrarily assigned, but they are internally consistent and perfectly valid for illustrative purposes. A going business can readily obtain actual costs from its accounting system to use in an analysis of this sort.[10] The refinements may provide more precise results; however the general conclusions will remain unchanged.

As shown by Table 10.1, there is a startling difference between the costs for complaints involving damaged containers, product quality, and product liability. The reason for such disparities can be explained by certain multiplier effects. In the case of the damaged containers, only 1 percent of the product needed to be replaced, but the quality complaint led to the return of the entire shipment. In the latter situation three freight bills were incurred instead of one,[11] and, on the assumption that the salvage value of the returned material was insignificant, the product cost was substantial. Equally dramatic, in the liability complaint 4 million pounds of the customer's product were ruined by only 40,000 pounds of ingredient. The cost of these failures escalate steeply from under $1,000 for a packaging problem to just under $25,000 for a product quality deviation, and over $1 million for a product liability.

In attacking the costs of customer complaints, management should allocate its resources where they will do the most good. There is no question that the incidence of damaged containers can be reduced by switching to stronger albeit more expensive packaging materials. After some improvements are made, however, a point of diminishing returns will be reached. Such a trade-off is inappropriate for liability complaints which must be dropped to an irreducible minimum. By spending more funds for prevention and for on-going testing, the cost of failure can be lowered.[12] Although the indirect costs of complaints have not been addressed in this discussion, such an omission is not meant to belittle the importance of customer good will and brand loyalty.

REFERENCES

1. "Will Send Consumer Complaints to Industry," *Food Technology*, April, 1984, p. 49.
2. "Procedure for Handling Shipping and Quality Complaints," Correspondence from U.S. Borax, Los Angeles, California, 1978.
3. Sales literature from Impact-O-Graph Division, Chatsworth Data Corporation, Chatsworth, California, 1980.
4. "Indicator Detects Thawing," *Chemical & Engineering News*, February 20, 1984, p. 31.
5. Ann C. Chadwick, *A Manual for Successful Resolution of Consumer Complaints in the Food Industry*, The Food Processors Institute, Washington, D.C., 1983.
6. *Information Master*, High Technology Software Products, Inc., Oklahoma City, Oklahoma, 1979.
7. Davis M. Woodruff and Felix M. Phillips, "The Pareto Chart . . . Tool for Problem Solving," *Chemical Engineering*, April 14, 1986, pp. 111, 112, 114.
8. *Statistical Quality Control Handbook*, Western Electric Co., Inc., Indianapolis, Indiana, 1956, pp. 144-148.
9. Sidney R. Daniel, "How to Develop a Customer Complaint Feedback System," *Food Technology*, September, 1984, pp. 41-46.
10. "Cost of Quality," *Quality Matters*, February, 1981, pp. 1, 7.
11. Orville R. Mertz, "Quality's Role in ROI," *Quality Progress*, October, 1977, pp. 14-18.
12. R. Morgan Anderson, "Controlling Food Plant Quality Costs," *Food Technology*, April, 1984, pp. 111, 112.

PART IV

Distribution

CHAPTER ELEVEN
PACKAGING

The distribution of safe and wholesome food products depends on the same principles of quality assurance as are required in manufacturing. For example, sanitation and personal hygiene, both so important in food processing, are sine qua non in distribution. Where manufacturing and distribution do diverge is in the fact that many more businesses and workers are directly involved in handling food products after the goods leave the plant. This dispersion of responsibility greatly complicates communication and supervision. While food manufacturing in the United States is concentrated in the hands of a few hundred producers, hundreds of thousands of retail food stores, restaurants, wholesalers, and distributors are positioned between the supplier and the ultimate consumer. To supervise this far-flung network, an elaborate infrastructure of federal, state, and local regulatory agencies has evolved.

Packaging can be thought of as the last step in manufacturing or as the commencement of the distribution chain. Because the packaging needs for a food product are so dependent on the manner in which it will be shipped and eventually used, there is ample logic for considering packaging as an adjunct of distribution. This association of packaging with distribution also underlines the extreme importance of packaging in quality assurance. Superior packaging provides the only insurance that what the customer receives will be the same as what the manufacturer ships.

The packaging engineer must have a detailed knowledge of the applications for which a food item is designed. His job is compounded on the one hand by the vast array of materials and technologies at his disposal and on the other by incessant consumer demands for new products. A summary of the functions which packaging is asked to perform will help to orient the persons responsible for packaging. Packaging may be used to —

contain such difficult to hold products as carbonated beverages;
preserve the freshness and integrity of foods throughout storage and distribution;
protect foods from environmental hazards as well as dangers inflicted by man;

insure against loss and pilferage during shipping and distribution;

unitize individual items by combining them into larger parcels;

communicate a message to the consumer that will assist or motivate him;

heat and serve meals prepared in advance, such as TV dinners;

apportion product into convenient quantities;

dispense foods without the need to transfer them to canisters, shakers, jars, or decanters;

double as reusable containers, utensils, or decorative pieces.[1]

Short of writing a lengthy treatise on packaging, only a limited number of topics could be touched upon. The packaging triumphs of canning and freezing have been critically reviewed in Chapters Two and Five. These technologies have been thoroughly developed and are now well ingrained in food science. By contrast, the subjects which follow here have been picked for their current interest and controversy.

TAMPER-EVIDENT CLOSURES

The murder of seven innocent people in 1982 marked the watershed of the packaging industry. This odious crime led to the enactment of legislation to curb and punish any acts of tampering with consumer products, including all food, drugs, and cosmetics. Of equal importance, the need for more secure closures for packaging was recognized by manufacturers. An observer of the scene, the executive editor of *Food and Drug Packaging*, predicted, "Packages designed to show evidence of tampering will be the wave of the future."[2]

The fateful event began to unfold on September 30, 1982, when three deaths, all attributed to cyanide poisoning, were reported. Upon taking a capsule of Extra Strength Tylenol to relieve a minor chest pain, a man living in the Chicago suburb of Arlington Heights became ill and soon thereafter died in the local hospital. That evening his younger brother and sister-in-law, while grieving for their deceased relative, each consumed medication from the same bottle of Tylenol. The brother died almost instantly, and his wife succumbed two days later. By the end of the week four more persons, living in communities outside Chicago, but all strangers, were dead from the same cause. Among the victims were a twelve-year old girl, a mother of four young children, and an airline stewardess in her thirties.

The connection between the deaths and Tylenol was made by two off-duty firemen. Upon hearing news reports of the tragedy, they arrived at a hunch which they passed on to officials investigating the case. A careful inspection of the partly used Tylenol bottles revealed that someone had tampered with the capsules. The red halves of the capsules were discolored, and when they were pulled apart, the capsules emitted the telltale almond odor of cyanide. An analysis confirmed that the Tylenol capsules contained cyanide far in excess of the lethal dose.[3] Every indication pointed to the fact that bottles of laced Tylenol had been planted in drugstores and supermarkets in a strip along Illinois Route 53 west of Chicago.

At this point, the investigation was stymied by the lack of any clues or motives. Billed as "Murder by remote control," the impersonal nature of the crime struck fear into the hearts of consumers.[4] A near panic gripped the country from coast to coast. Leading psychologists and psychiatrists were consulted, but other than typing the criminal as a paranoid or a psychotic, these professionals were stumped. Their only prognosis was extremely unsettling: other mentally deranged individuals might be tempted to "copycat" the Tylenol crime very much as hijackings had previously swept commercial aviation. (As a sequel to the investigation, a newspaper account more than two years later reported that the cause of the deaths remained a mystery, and the culprit, as far as anyone could determine, still remained at large. The Tylenol episode most likely would go down in the annals of crime as unsolved.)

The reaction of the manufacturer of Tylenol was swift. Johnson & Johnson, whose subsidiary, McNeil Consumer Products, produces the painreliever, immediately undertook a total recall of Tylenol capsules throughout the United States. Eventually this action cost the company $50 million net after taxes. Management then made a bold decision. It confounded the pundits by opting to stick with the name, Tylenol, and to reintroduce the product in a new tamper-resistant package. This move not only was a turning point for Johnson & Johnson but for the entire industry as well. The wisdom of this decision was later proved when the new Tylenol product succeeded in all but recapturing its lost market share, notwithstanding stiff competition.[5]

Government Response

From the beginning of the Tylenol incident, government regulators were stunned and perplexed about what to do. Their reaction, however, was to do something, almost anything, to shore up the confidence of the public. The Cook County Board of Commissions immediately passed an ordinance that required seals on all nonprescription drugs. Food and Drug Administration (FDA) Commissioner Arthur H. Hayes, Jr., however, was not reassuring. He pointed out that "it is important to make clear that a tamperproof package is not possible."[6] His assessment of the problem did not prevent FDA from submitting in record time on November 5, 1982, a final regulation requiring tamper-resistant packaging for over-the-counter (OTC) drug products.[7] These regulations preempted state and local requirements in recognition of the fact that practically all these products are sold in interstate commerce.

The new FDA regulations covering tamper-resistant packaging conspicuously exempted foods. Dr. Hayes again was quoted as saying,

> I'm not going to write regulations for foods or anything else until I see a need. I promise you — that if ever the time comes that regulations for packaging or labeling or displaying in the food industry or any other industry under our purview [are needed] to achieve an agreed-upon or stated goal — we will do it! But I'm going to wait until we know what the goal is and indeed what the *problem* is that allows us to determine the goal.[8]

In fact, many food products were already being packaged in tamper-resistant containers. As far back as 1977 proposed revisions in the Pasteurized Milk Ordinance required tamperproof caps and closures for containers of milk and milk products.[9] Under the impetus of the Tylenol crisis, the food industry was moving with alacrity to convert more products to safer packaging. This effort would have progressed even faster save for the complexity and diversity of food packaging problems which defied the adoption of simplistic solutions.

Even though food products were not covered by FDA regulations for tamper-resistant packaging, they were included in new legislation passed by Congress in 1983. The Federal Anti-Tampering Act (Public Law 98-127) was enacted making the tampering or tainting of any consumer product a criminal offense. The law provides for fines and prison terms for violations involving all food, drug, and cosmetic products. In such cases the Federal Bureau of Investigation (FBI) is given overall jurisdiction while FDA and the U.S. Department of Agriculture (USDA) may be assigned joint investigative powers as needed.[10]

Designs of Tamper-Evident Closures

Realizing that there is no such thing as a tamperproof package, the food industry has turned to the concept of "tamper-resistant" or more aptly expressed "tamper-evident" packaging. The possible designs are limited only by the imagination of the packaging engineer. Experts consider the safest packages to be those equipped with seals or other devices that the user must break to gain access to the product. A brief run-down of the most popular tamper-evident closures illustrates their features.

• Blister packs are in ever increasing demand for pharmaceutical products in tablet or capsule form. The pills are inserted in bubbles formed between plastic laminations, which must be torn to remove the product. While not applicable for most food products, blister packs nonetheless can be used for vitamin supplements, seasonings, and other individualized food items.

• Aluminum foil and waxed-paper seals covering the mouths of jars have been used for some time to protect the freshness of instant coffee and vitamins. The seal, which is stuck to the glass rim by adhesive, must be punctured to remove the contents.

• Heat shrinkable plastic bands are easy to apply over screw caps or other bottle closures. A number of food companies, in a haste to obtain some protection from tampering, have turned to this method. One producer was able to convert his filling line for steak sauce in a matter of days without the need to redesign his basic package. The band, which may be perforated, can be printed with the manufacturer's logo as an added attraction and also as insurance against substitution. To open the container, the band must first be ripped off or split at the perforation.[11]

• Screw caps with breakaway rings are becoming more common. When the cap is twisted, it separates from the ring, which remains on the bottle. Such caps are available in aluminum or plastics.[12]

• Lids with built-in safety buttons have been used since the early 1960's to close glass jars of baby food. As long as a vacuum is maintained in the jar, the button remains depressed. The "big dimple" provides visual evidence that the jar has been properly sealed and a shopper has not meddled with it. When the customer opens the jar, she should hear a pop which gives added assurance that the container was properly closed.[13]

Tamper-evident closures provide assurance not only against the malicious acts of criminals but also the inconsiderate impulses of shoppers. People somehow cannot resist their urges to inspect, smell, taste, or poke the food that they and others will buy. Before babyfood manufacturers converted their packaging to the new lids with safety buttons, mothers were occasionally caught "sampling" jars sitting on supermarket shelves.[14] In this day of self-service retailing, such human behavior is difficult to police. Therefore any added protection that packaging improvements offer should be exploited by food manufacturers.

NEW TRENDS IN CONSUMER PACKAGING

Although the issue of tamper-evident packaging has overshadowed the food industry since the Tylenol affair, other notable trends deserve mentioning. These developments depend on the utilization of emerging technologies and materials that were nonexistent a short time ago. Such innovations promise benefits to the consumer in the way of quality, convenience, safety, and economy. Every food technologist has his own favorite list of outstanding packaging achievements; however, the following examples would probably appear on most of them: aseptic packs, retort pouches, microwavable trays, squeezable tubes, and plastic bottles.

Aseptic Packaging

A trend, little noticed by Americans, has been the steady conversion of the dairy industry in Western Europe to aseptic packaging. By 1984 the market inroads made by this new technology was estimated at 40 percent. Possessing a five month shelf life when stored at ambient temperatures, aseptically packaged dairy products have benefited lower income groups that do not have modern appliances. No longer do European households without refrigerators have to boil their milk supplies and take other added safety precautions.[15]

So far, aseptic packaging is being used in the United States only for high-acid foods. In 1984 about 5 percent of the fruit juice and 20 percent of the fruit drink volume moved in aseptic packs. These "paper" boxes are fabricated from laminates containing aluminum foil sandwiched between layers of plastic film and paperboard. Complete with a drinking straw, the 250 ml. aseptic box is popular for "brown bags" (which have replaced lunch pails) and snacks away from home.[16]

The introduction of aseptically packaged low-acid foods including milk products is lagging in the United States. One reason is the prevalence of refrigeration. The other factor delaying progress is the need to master extremely sophisticated technology. In the aseptic packaging of low-acid foods, spores of *Clostridium botulinum* must be completely destroyed. This need is tantamount to requiring that the product, package, and filling environment be sterile. The use of hydrogen peroxide has been a great help in effectively sterilizing containers; however, the condition of the filling equipment is more difficult to control.[17,18] A final nod of approval will be required from FDA before producers are allowed to market aseptically packaged milk in the dry goods section of supermarkets. Closest to receiving approval are flavored milk drinks which very likely will attract the greatest interest from consumers.[19]

Retort Pouch

The retort pouch preserves food very much like tin cans but offers several advantages over cans. Retorting times are shorter for pouches, thus achieving energy savings. The less severe heating step also keeps flavors and nutrients better. Retort pouches are less energy-intensive than frozen foods, which must be held in freezers until they are consumed. Low-acid particulate products, such as beef stew, can be stored in retort pouches for up to a year and a half on the cupboard shelf. When the housewife is ready to prepare a meal, all she does is drop the pouch in boiling water for five minutes and empty the contents on a plate without sullying a pot.

In spite of the several advantages of the retort pouch, FDA was slow to approve it. Even though the film employed to form the pouch did not present any safety problems, the agency was concerned about the possible migration of the polyurethane-epoxy glue used to seal it. The film is made of an exterior ply of polyester plastic, a middle layer of aluminum foil, and polypropylene on the inside. This film is considerably more expensive than tin cans. Still, the advantages of the retort pouch were so significant that the military pressured FDA to approve it in 1978. Besides offering superior quality and saving energy, the pouch is light in weight and easy to stack in boxes. While food manufacturers have been slow to exploit the retort pouch in commercial applications, its future appears bright.[20]

Microwave Trays

With the advent of microwave ovens, a need has arisen for compatible packaging materials. By 1987 an estimated 60 percent of American homes were equipped with microwave ovens, and Pillsbury Co. believes that proportion will reach 90 percent by 1990.[21] In order to make packaging "microwaveable," suitable plastics and paper trays have to be substituted for metal films. Food products so packaged can be heated directly in their wrapping for fast and convenient service.[22]

Tubes

Squeezable tubes have largely escaped the attention of American food manufacturers. In Europe, all sorts of condiments, patés, spreads, and icings are conveniently and safely packaged in these containers. Fitted with special openings, these tubes can form fluted or other fancy squirts to decorate canapés, hors d'oeuvres, appetizers, salads, desserts, and other dishes. The Tube Council of North America is promoting these containers as being sanitary, economical, and nonbreakable. Its efforts could turn out to be quite successful.[23]

Plastic Bottles

Plastic bottles, which possess many remarkable properties, have been held up in their development by unanticipated legal hurdles. As is the case with all food packaging materials, any plastic proposed for making bottles must be exhaustively tested and found to be unquestionably safe before receiving FDA approval. Section 201(s) of the Federal Food, Drug & Cosmetic Act defines food additives as including any substance used in packaging that indirectly may become a component of food unless it is generally recognized as safe. Thus, packaging materials come under all the provisions for food additives, including the so-called Delaney Clause, which strictly prohibits any compound suspected as being a carcinogen.

FDA regulations require:

> Packaging processes and materials shall not transmit contaminants or objectionable substances to the products, shall conform to any applicable food additive regulation (Parts 170 through 189 of this chapter), and should provide adequate protection from contamination. (21CFR110.80 [h])

Consistent with FDA regulations, USDA policy specifies that all meat and poultry plants must obtain from their suppliers and keep on file written guaranties for all packaging materials. These guaranties are to vouch that such wrappers and containers comply with the federal food laws and regulations.[24] One area where both USDA and FDA are less clear concerns colorants and printing inks. A limited list of approved pigments was published in 1972, but afterwards newer products were reviewed on an individual basis.[25] Then in 1983 FDA issued a final rule which backtracked on its previous rulings but failed to provide an updated list of pigments. Until new directives are forthcoming, packaging suppliers will be operating in a state of limbo.[26]

In order to comply with federal regulations, extraction testing must be conducted on all packaging materials. This work is done to determine the possibility of a packaging component migrating into and becoming part of the food product. Designing a proper experiment requires the careful consideration of actual conditions, such as the holding time, temperature, surface area, and food properties. Two approaches are acceptable to FDA: either long term storage studies with the given food may be performed, or extractive tests with solutions that simulate true conditions may be used. FDA has suggested that the following solutions be considered for extractive studies:

1. distilled water,
2. 10 percent and 50 percent alcohol,
3. corn oil,
4. sodium bicarbonate,
5. 3 percent acetic acid, and
6. 20 percent sucrose.[27]

Extended studies on migration by the consulting firm, Arthur D. Little, have indicated that more representative results are obtained by substituting 8 percent aqueous ethanol for water, and by using 95 percent ethanol instead of heptane when corn oil cannot be used as a fatty food simulant on olefin contact surfaces. To evaluate leaching from polystyrene articles by fatty foods, a 50 percent ethanol solution is recommended.[28,29] Because of the extended shelf-life of most products, accelerated testing at elevated temperatures is more practical. Petitions submitted to FDA must identify the composition of the packaging component, give the amount extracted, and supply toxicity data on it.

In practice, the regulatory procedure for the approval of packaging can result in costly delays and setbacks. A classic case was the petition filed by Monsanto Company to use acrylonitrile-styrene copolymer resin in non-alcoholic beverage bottles. After receiving a provisional approval in 1974 from FDA, Monsanto built three blow-molding facilities to handle a major contract from The Coca-Cola Company. No sooner were these plants in production than FDA changed its mind in the light of new evidence that acrylonitrile was a suspected carcinogen and some unreacted monomer could migrate into the product. With 21 million useless bottles on its hand, Monsanto sued the government but lost the case.[30] In 1984, ten years after initiating the project, the company finally received a permit to use acrylonitrile copolymers in bottles. This reversal, though, was based on substantial improvements in the resin whereby residual monomer has been reduced by more than 100-fold.[31]

In the meantime other plastics have been certified for beverage bottles. One of the outstanding resins is polyethylene terephthalate or PET for short. Excellent consumer acceptance of PET soft-drink bottles has forced resin manufacturers to expand capacity. Plastic bottles that are impermeable to oxygen have been produced from co-extruded plastic sheet containing polypropylene and a barrier film of ethyl-vinyl alcohol copolymer (EVOH). These containers, which are tough and light, are being touted for such oxygen-sensitive foods as ketchup, barbecue sauce, mayonnaise, and salad dressings.[32,33]

Many of the newer packaging materials are finding their way into kitchenware. Small plastic tubs used to hold leftovers are little different from many food containers. For this reason, FDA has reexamined its stance on these articles. When the Food Additives Amendment was passed in 1958, Congress had no intention of regulating housewares, which consisted mostly of pots, pans, and dishes. These assumptions, according to FDA, are no longer valid.[34]

INDUSTRIAL PACKAGING

Multiwall Bags

Because multiwall bags are so versatile and economical, they are the standard means of shipping food ingredients in less than bulk load quantities. These packages are available in a variety of sizes, typical net weights being 50 and 100 pounds. They can be adapted to hold many kinds of granular and powdered products and to meet diverse shipping conditions. Above all, they are inexpensive and disposable, thus providing convenience to both supplier and customer.

Multiwall bags are constructed of two to six plies of kraft paper or plastic film. These layers are unstuck from each other, thereby providing much greater flexibility and strength than one sheet of equivalent total thickness would give. The gauge of each ply is specified by giving its basis weight, which is defined as the weight in pounds of 3,000 square feet of paper or a stack of 500 sheets each 24 by 36 inches. Common weights of kraft are 40, 50, 60, and 70 pounds. Extensible kraft paper may be substituted for paper of standard strengths to provide greater resilience or to reduce the weight of the bag.[35]

Barrier properties may be provided by including a film of polyethylene (PE) in the bag construction. This film, commonly a 3 mil thickness of low-density PE, usually is placed on the inside. It can protect the product from the pickup of moisture or objectionable odors. It also provides resistance to grease. If the food product is hygroscopic or oily, a PE film is important for package integrity.

The outer ply of a multiwall bag may be treated in a number of ways to provide functional properties. Many food manufacturers specify a bleached-white outer wrapping to differentiate their products from non-food grade materials, although this practice is by no means universal throughout the industry. A silicone treatment may be specified to provide water repellency, particularly in the case of hygroscopic products which may adhere to the outside of the bag and cause stains. Anti-skid finishes help in palletizing bags as is discussed later in this chapter.[36] Various coatings are available to resist abrasion and scuffing. Finally, a pesticide such as pyrethrin may be applied to give insect resistance, especially to bags used to ship commodities overseas under U.S. Government programs.[37]

There are two major bag designs although numerous variations of these types exist. The open-mouth (OM) bag comes sealed at one end and open at the other for filling. Once it is filled, it is closed either by pasting or sewing. The stitching of a sewn bag may be covered by tape to make the closure sift-proof or to protect against moisture or vapors.

The second basic bag design is the valve bag. This bag is completely closed except for a small sleeve-like opening at the top of one side. To fill the bag, the sleeve is slipped over a horizontal filling spout. When the bag is full, the sleeve is tucked inside where it is held shut by the pressure of the bag's contents. In order to increase the volume of either the OM or valve bag and also to square the sides for improved stacking, gussets that look like pleats are built into the sides. Other design options include a choice of either a pinch bottom or satchel bottom.[38] Bag manufacturers have expended a great deal of effort to optimize design.

Of late, the pasted open mouth bag has received the most attention from food producers. If properly sealed it provides a positive closure that keeps contaminants from entering the bag and prevents product from leaking out. For added protection a double seal may be applied by first heat sealing the inner plastic ply. The kraft paper plies are closed by folding the top lip over and heating the precoated hot melt adhesive. Because the multiple plies are stepped, a positive seal is realized for each ply. This bag is designed to give a rugged and tight closure that will endure rough handling.[39]

Under actual operating conditions the pasted open mouth bag too often delivers less than perfect results. If the hot melt adhesive has not been evenly applied or if it is not sufficiently heated in the packaging line, a poor seal will result. Or if the product is hot when being packaged, it may prevent the adhesive from cooling rapidly enough to stick before the bag is released. In case the bag is misaligned or the speeds of the conveyor belt and the sealing head are unsynchronized, the closure will be skewed and may fail at the corners. Powdered materials can also cause problems. Even though deaerators are available for packaging powdered materials, excessive dusting may interfere with a tight seal.

The valve bag still enjoys considerable popularity in the food industry, particularly for less sensitive products such as starch and sugar. Design improvements in the valve bag have increased productivity and at the same time minimized leakage thereby making this package more attractive. In one modification, a PE film inserted in the valve acts as a flap to reduce sifting.[40] Another proposal is to apply a tape over the valve inlet after filling the bag in order to gain a tight closure.

Stringent specifications have been established for multiwall bags. Packaging must comply with the rules of the Uniform Freight Classification for rail shipments and with the rules of the National Motor Freight Classification for truck shipments. These rules include standards of construction as well as performance tests, such as the drop test for impact resistance. New packaging must be tested in trial shipments before being approved for either rail or highway movements.[41] To be in compliance, each bag must be marked with a freight classification stamp indicating the name of the bag manufacturer. Other container specifications may be promulgated by USDA for shipping various commodities.[42] Packaging standards covering construction, performance, and quality are critical in determining negligence in product liability lawsuits.[43,44]

Fiber Drums

Fiber drums, although considerably more expensive than multiwall bags, may be justified for packaging such high value products as vitamin blends. One or more of the following advantages may dictate their use:

1. Drums provide better protection of fragile products than do bags.
2. With drums there is reduced loss of product through damage.
3. Because drums are reclosable, multiple withdrawals of product are facilitated.
4. Empty drums may be reused for other products or applications.
5. Export orders may specify drums because of their durability.
6. In limited storage space, drums can be stacked several tiers high.

The strength of a fiber drum is related to the thickness of its cylindrical shell. Depending on needs, the bursting strength can be varied from 250 to 900 pounds per square inch. The tops and bottoms may be fabricated from steel or fiber board. Steel lids can be securely locked and sealed to discourage tampering. Alternatively, telescoping fiber tops, which fit over the shell, may be secured with pressure sensitive tape. Usually a liner of PE film is placed inside the drum and sealed with a plastic tie. For improved weatherability, an external finish of varnish or paint may be applied to the shell. Fiber drums come in a wide range of sizes and can easily hold up to 400 or more pounds of product.

Because of the durability and serviceability of fiber drums, there is a great temptation to reuse them, sometimes improperly. If they are refilled with food products, extreme care must be exercised to ascertain their cleanliness. FDA, mindful of potential problems of using secondhand containers, warns that "one person's trash is another's treasure."[45] The food manufacturer is held responsible for knowing the condition of all the containers it uses.

Semi-Bulk Containers

Semi-bulk containers, each with a capacity of 2000 pounds or greater, have been employed for some time in special circumstances. Cubical metal bins are typical of this kind of packaging, which is often used in the plant for work-in-progress.[46] A newer configuration consists of a polypropylene fiber bag fitted with a PE inner liner. These semi-bulk sacks are slung by straps in order to move them by fork lift trucks.[47]

The semi-bulk sack has been promoted as a labor saving device as well as a means of economizing on packaging costs. Maximum savings are realized by utilizing returnable bags that can make at least several round trips. Studies, however, have suggested that the potential for contamination in this mode of operation is too great and that semi-bulk bags should be restricted to one-way use. A further drawback to this type of packaging is the need to install special filling and unloading devices. Not to be overlooked is the possibility of puncturing a semi-bulk bag in which case there is a much bigger mess to clean up than with a multiwall bag. Greater use of semi-bulk containers will probably be limited not only by the above-mentioned disadvantages but also by the outstanding success in unitizing smaller containers.

UNITIZING

Unitizing can be defined as the bundling together of smaller packages into larger self-contained units. Many examples exist of unitizing in the food industry. Packs of chewing gum may hold five or so sticks of individually wrapped gum. Nowadays beer and soft drinks are sold by the sixpack. On a larger scale, forty fifty-pound bags may be stacked on a mobile wooden platform, roughly four feet square, known as a pallet. Palletized in this manner, the bags can be moved as a single unit from the manufacturer to an intermediate public warehouse and then to their final destination. For all practical purposes the package assumes the attributes of a 2000-pound container.[48]

Palletizing offers to food manufacturer several advantages. First, by combining smaller packages into larger units it saves significant labor costs. Second, by applying an outer wrapping, added protection against contamination is assured. And third, pilferage and loss are more easily spotted and controlled. Because efficient materials handling is aided by palletizing, this practice should continue to expand. The increased use of palletizing will also bring undeniable benefits to quality assurance by reducing the need for manual operations.

A high degree of automation is possible in palletizing multiwall bags. In a modern layout, a conveyor leads the bags away from the filling line to a flattener which redistributes the product within each bag. The bags then pass over a checkweigher and through a metal detector. More will be said later about each one of these instruments. An automatic palletizer loads the bags onto pallets according to a predetermined pattern. To provide improved stability, the stacked bags may be tacked together by a daub of adhesive on each bag. A cover of heat-shrinkable film is slipped over the entire palletized load, which is then passed through a shrink tunnel.

Some modifications in palletizing are possible. To save some expense, shrink wrapping may be replaced by stretch wrapping; however, the top of the load is left unprotected by this method. Another cost-saving substitution is the use of slipsheets instead of pallets. Cut to the same size as the pallet, slipsheets are available in different materials: corrugated paper, fiberboard, and plastic. Although slipsheets require special handling equipment and greater operator skill, they take less room and are disposable.

Even when using pallets, it is considered good practice to use a slipsheet under the bottom row of bags to protect them from wooden splinters and protruding nails. A slipsheet will also offer some protection against inadvertent pallet contamination or insect infestation. When stacking palletized loads on top of each other, dividers, similar to slipsheets, should be placed on the upper tier of bags. These sheets will keep the shrink film from being scuffed and torn. All pallets, particularly returned ones, should be meticulously inspected to make sure they are sanitary and free from harmful substances. Some shippers prefer the alternate use of pallets and slipsheets. In this fashion, pallets, which remain captive within a facility, are used to move product around the plant or warehouse, and slipsheets are employed to transport unitized containers by truck or rail.[49,50,51,52]

Extensive precautions are required in placing palletized loads on cars or trailers. Without adequate bracing and the use of dunnage, loads will shift during transit causing extensive damage and requiring extra labor for unloading. Several effective and economical products are available to use as dunnage. Inflatable bags made of high-strength kraft paper can absorb substantial shock.[53,54,55] Paper honeycomb also is used to fill voids between pallets.[56,57] A shipper needs to witness only once a disaster at the receiving end of a shipment to be convinced of the need for proper bracing and dunnage.

Beyond the benefits already mentioned, unitizing has the advantage of reducing record keeping and simplifying quality control. To realize this plus, however, only one code date (lot number) can be included in each unitized load. Regardless of circumstances, replacements should not be made from another code date. For example, if a bag is damaged in a palletized load, it should be replaced by another bag from the same code date. Should a substitute not be available, the pallet ought to be shipped short one bag and so documented. Nothing can be more disconcerting to a customer than receiving a scramble of code dates in a palletized load or, for that matter, within a carload shipment.

PACKAGED WEIGHT CONTROL

A discussion of net weight control provides an excellent opportunity to introduce the fundamentals of statistics as related to quality assurance. This mathematical discipline is vital for understanding the significance of large quantities of data. Statistical theory also is absolutely essential when tight control is desired over a critical variable. Instead of packaged net weight, that variable could very well be product purity or pH. Since the same principles apply regardless of the given variable, this treatment of net weight control is relevant to all quality control functions.

Fundamentals of Statistics

No one should be surprised to find when weighing many objects, each of which is produced in an identical process, that few, if any, of the results will be the same. Instead, the observed values will be clustered around some average weight. Furthermore, if the objects are produced in a haphazard manner or if the measurements are crude, the results will be more scattered than would otherwise be expected. To visualize the variation in the results, the data can be presented as a kind of bar chart, more particularly known as a histogram.[58] The width of each bar is proportional to an incremental weight while the height of a particular bar represents the frequency in terms of the number of objects within the given increment. Results plotted in this way will look like the graph in Figure 11.1.

The more objects weighed from a given source, the greater will be the accuracy of the results. Again, these measurements may be depicted by a histogram, this time containing more bars, all of which are thinner. Eventually, as the number of objects weighed approaches infinity, the graph takes on the appearance of a smooth bell-shaped curve, similar to the one in Figure 11.2. This curve, which is closely approximated by many types of variables, is known as a normal frequency distribution.[59]

Several characteristics of a normal frequency distribution are important in statistics. The peak of the curve will be located at the distribution or population mean, μ (mu). In a limited number of measurements, μ can be estimated by the sample mean or average, $\bar{x}$, provided that testing methodology is correct, i. e., there are no biases in the system.[60] To calculate $\bar{x}$ the following equation is used, where x is the variable, e.g., weight, and n is the number of objects measured:

$$\bar{x} = \frac{\Sigma x}{n} \quad [61]$$

The spread of a normal frequency distribution is specified by its standard deviation, represented by the symbol σ (sigma). It too can be estimated from a limited number of data by calculating the sample standard deviation, s, as follows:

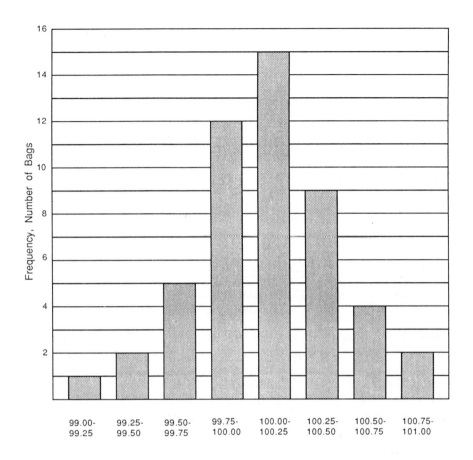

FIGURE 11.1. HISTOGRAM OF NET WEIGHTS OF BAGS

$$s = \sqrt{\frac{\Sigma(x - \bar{x})^2}{n - 1}} \quad \text{or}$$

$$s = \sqrt{\frac{\Sigma(x^2) - n\bar{x}^2}{n - 1}} \quad [62]$$

The latter expression is an equivalent form which is easier to calculate. Sigma values are extremely useful in determining the distribution of data. Thus, for a normal frequency distribution, about 68 percent of all the data will lie between $\pm 1 \, \sigma$ of the mean, 95 percent within $\pm 2 \, \sigma$, and 99.7 percent will be bounded by $\pm 3 \, \sigma$.

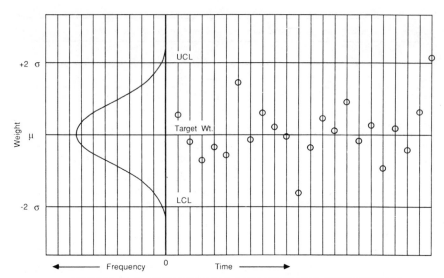

FIGURE 11.2. THE NORMAL FREQUENCY DISTRIBUTION
AND A CONTROL CHART

Not all data closely fit a normal frequency distribution. Many variables only approximate it, and others in no way resemble it. In these situations, calculated values of sigma give erroneous results. Fortunately, data can readily be transformed to achieve normality. The Central Limit Theorem states that averages, each calculated from a given number, n, of independent measurements, more closely approximate normality than the separate measurements. As n increases, the fit becomes better.[63] Often n is set at 5 because of the ease of calculation. The average of five numbers can be obtained by multiplying the sum by two and moving the decimal point one place to the left.[64]

The Control Chart

Since its development in 1939, the control chart has been one of the most useful tools in quality control. It can be thought of as a frequency distribution curve but turned on its side. The vertical axis indicates weight while the horizontal axis has been modified to represent a time function instead of frequency. (Figure 11.2) The average or mean of the normal frequency distribution becomes the target weight of the control chart. An upper control limit (UCL) and a lower control limit (LCL) are established at some fixed values.

For a majority of quality control applications the UCL and LCL are set at ± 2 sigma respectively.[65,66] Thus, if the target weight is put at 2 sigma above the packaged label weight, over 97.5 percent of the containers will exceed the specification. While such a situation is generally regarded as acceptable by consumers, the manufacturer will be penalized by giving away product equal to the difference between the target weight and the label weight. For this reason, industry is provided a strong incentive to improve its control over packaged weight so as to reduce the value of sigma.[67] The smaller sigma is, the closer the target weight can be brought to the label weight. Over the long term a manufacturer cannot afford to operate inefficiently: it must minimize overfill either by reducing the target weight or by increasing the label weight and charging a higher price.

Food manufacturers may achieve additional economies by establishing the target weight at less than 2 sigmas above the label weight. Just how far they can adjust their fill in order to save product is specified by government regulations. Two agencies, FDA and the National Bureau of Standards (NBS) of the U.S. Department of Commerce, have concurrent jurisdiction over the control of the weights of packaged food products. In the case of meat and poultry products, FDA is superseded by USDA. All of the agencies recognize and make allowances for statistical variations in packaged weight.

National Bureau of Standards

The NBS regulations were first spelled out in Handbook 67, issued in 1959. This publication was updated in 1981 as Handbook 133, *Checking the Net Contents of Packaged Goods*.[68] As adopted by the National Conference on Weights and Measures, it serves as a model regulation for adoption by the states, more than forty of which have passed enabling legislation.[69] Handbook 133 sets forth two principles for the control of weights of mass produced packaged goods:

1. The average quantity of contents of a lot, shipment or delivery of packages must at least equal the quantity printed on the label.
2. The variation of individual package contents from the labeled quantity must not be "unreasonably large."

The Handbook goes on to define "unreasonably large" in terms of Maximum Allowable Variations (MAV), which are given in Figure 11.3. For packages over 2 pounds, the MAV is 1 percent. This allowance is increased to 3 percent for smaller units, sizes greater than 2 ounces and up to and including 2 pounds. The allowance for packages weighing 2 ounces or less is 5 percent. These MAV's are used as standards to control minus deviations only. Any positive variations are left to the competitive pressures of the marketplace to regulate.

FDA's Policy

FDA's policy on net contents is outlined in its regulations as follows:

> The declaration of net quantity of contents shall express an accurate statement of the quantity of contents of the package. Reasonable variations caused by loss or gain of moisture during the course of good distribution practice

or by unavoidable deviations in good manufacturing practice will be recognized. Variations from stated quantity of contents shall not be unreasonably large. (21CFR101.105 [q]).

Maximum allowable variations for an individual package labeled by weight.[a]

Avoirdupois units			Metric units	
Labeled weight	MAV		Labeled weight	MAV
Pounds or ounces	Decimal pounds	Fractional ounces	Grams	Grams
up to and including 0.026 lb up to and including 0.41 oz	0.001		up to and including 0 to 11.6	0.5
0.026+[b] to[c] 0.04 lb 0.041+ to 0.64 oz	0.002	1/32	11.6+ to 18	1
0.04+ to 0.08 lb 0.64+ to 1.28 oz	0.004	1/16	18+ to 36	2
0.08+ to 0.12 lb 1.28+ to 1.92 oz	0.008	1/8	36+ to 54	4
0.12+ to 0.18 lb 1.92+ to 2.88 oz	0.012	3/16	54+ to 82	5
0.18+ to 0.26 lb 2.88+ to 4.16 oz	0.016	1/4	82+ to 118	7
0.26+ to 0.34 lb 4.16+ to 5.44 oz	0.020	5/16	118+ to 154	9
0.34+ to 0.46 lb 5.44 + to 7.36 oz	0.024	3/8	154+ to 209	11
0.46+ to 0.58 lb 7.36+ to 9.28 oz	0.028	7/16	209+ to 263	13
0.58+ to 0.70 lb 9.28+ to 11.20 oz	0.032	1/2	263+ to 318	15
0.70+ to 0.84 lb 11.20+ to 13.44 oz	0.036	9/16	318+ to 381	16
0.84+ to 0.94 lb 13.44+ to 15.04 oz	0.040	5/8	381+ to 426	18
0.94+ to 1.08 lb 15.04+ to 17.28 oz	0.044	11/16	426+ to 490	20
1.08+ to 1.26 lb	0.048	3/4	490+ to 572	22
1.26+ to 1.40 lb	0.052	13/16	572+ to 635	24
1.40+ to 1.54 lb	0.056	7/8	635+ to 698	25
1.54+ to 1.70 lb	0.060	15/16	698+ to 771	27

[a]Applies only to shortages in package weight (minus package errors)
[b]0.026+ means "greater than 0.026".
[c]"to" means "to and including".

FIGURE 11.3.

Maximum allowable variations for an individual package labeled by weight.

Avoirdupois units			Metric units	
Labeled weight	MAV		Labeled weight	MAV
Pounds	Decimal pounds	Fractional ounces	Kilograms	Grams
1.70+ to 1.88	0.064	1	771+ to 852	29
1.88+ to 2.14	0.070	1 1/8	852+ to 971	32
2.14+ to 2.48	0.078	1.1/4	971+ to 1.125	35
2.48+ to 2.76	0.086	1 3/8	1.125+ to 1.350	40
2.76+ to 3.20	0.094	1 1/2	1.350+ to 1.600	45
3.20+ to 3.90	0.11	1 3/4	1.600+ to 1.800	50
3.90+ to 4.70	0.12	2	1.800+ to 2.100	55
4.70+ to 5.80	0.14	2 1/4	2.100+ to 2.640	65
5.80+ to 6.80	0.15	2 1/2	2.640+ to 3.080	70
6.80+ to 7.90	0.17	2 3/4	3.080+ to 3.800	80
7.90+ to 9.40	0.19	3	3.800+ to 4.400	85
9.40+ to 11.70	0.22	3 1/2	4.400+ to 5.200	100
11.70+ to 14.30	0.25	4	5.200+ to 6.800	115
14.30+ to 17.70	0.28	4 1/2	6.800+ to 8.20	130
17.70+ to 23.20	0.31	5	8.20+ to 10.60	145
23.20+ to 31.60	0.37	6	10.60+ to 14.30	170
31.60+ to 42.40	0.44	7	14.30+ to 19.25	200
42.40+ to 54.40	0.50	8	19.25+ to 24.70	230
54.40+	1%		24.70+	1%

FIGURE 11.3. (*continued*)

Source: *NBS Handbook 133*, U.S. Department of Commerce, Washington, DC 20234, June 1981.

This policy is enforced by the following directive given in FDA's *Administrative Guidelines Manual*, Section 7420.04. Legal action will be recommended if:

> The sample shows an average (48 units examined) short weight of one percent or more.
> *and*
> Shrinkage, or other loss in weight due to storage or shipping conditions, and variations in tare weights have been ruled out as factors contributing significantly to the apparent shortage.

Federal regulations on net contents have been contentious because they provide for loss of weight due to moisture fluctuations during distribution. At the heart of the issue, federal inspectors generally check weights at the manufacturer's level, whereas state officials, who rely mostly on retail inspections, may have difficulty in correlating their observations with declared quantities. In 1971, California ordered two companies, Rath Packing Co. and General Mills, to desist from selling bacon and flour products respectively which were short weight as a result of moisture loss.[70,71] The manufacturers appealed the case and after years of litigation, the U.S. Supreme Court ruled in favor of these companies.[72] Thereupon, several states attempted to get FDA and USDA to change their policies, but these efforts were to no avail. Thus, the principle allowing "reasonable variations caused by loss or gain of moisture" is left standing.

With the availability of modern checkweighing equipment, manufacturers no longer need to weigh manually sample containers and to plot laboriously the data on control charts. Automatic scales with microcomputers placed in the conveyor line can weigh each unit, compute the net weight, and calculate meaningful statistical information, such as average weight and standard deviation.[73,74] Increases in the standard deviation will quickly flag difficulties with the filler. Likewise, any drift in the average weight will require corrective action. An eject mechanism can be installed to catch any grossly underfilled containers. Printers are universally available to provide printouts of essential data. As is done with other quality control results, a prudent manufacturer will maintain a permanent record of all packaged weights.

METAL DETECTORS

The perfect stage for quality control observations is after the product has been packaged and sealed in containers. Assurance is thereby given that data so collected will reflect the true condition of the product as received by the customer. Unfortunately, with present-day technology, many properties cannot be tested at this point without the necessary destruction of the package. As indicated in the previous section, however, net weight can be monitored after the package is filled and sealed. Another attribute, the presence or absence of foreign objects, also can be effectively controlled by the use of metal detectors located ideally at a point following the packaging line.

Principles of Detection

Metal detectors come in all shapes and sizes. The oldest and simplest configuration is the bar magnet, which may be placed in a processing stream to pick up tramp metal. Periodical inspections are required to observe any contamination and to clean the magnet. While straightforward in concept, this method has drawbacks. If the operator forgets to look at the magnet, collected deposits, such as a clump of shavings, may shake loose and cause more problems by contaminating a single batch or container rather than being dispersed evenly throughout a large production run. Further, magnets are incapable of picking up non-magnetic materials like aluminum, brass, and most stainless steel alloys. Perhaps the most serious limitation of magnets is that they are useless in checking packaged products. While magnetic traps fall short of providing the protection called for, they will continue to have their place along with sieves in production systems.

Radio frequency (RF) generators have been incorporated in another type of metal detector. These instruments provide for the sensitive detection of all conductive materials, including metals regardless of magnetic properties. Although the circuitry of these instruments may seem complex, the principle of RF detectors is simple. As a package is passed through an electromagnetic field, small electrical currents will be induced in any conducting material. These eddy currents will cause a field distortion, which will be detected by the instrument. Sensitivity will depend on the size of the instrument's aperture through which the package travels. Smaller openings produce stronger fields and therefore result in more sensitive detection. The best instruments available can signal the presence of metal particles as small as 0.03 inches (0.76 mm.) in diameter when scanning small packs.[75]

Third generation detectors are appearing on the market, based on the use of x-rays to pick up not only metal contaminants but also pits, stones, glass splinters, and other foreign matter. The resolution of these instruments is good enough to detect particles 0.06 inches (1.5 mm.) in diameter.[76] This technology is an outgrowth of the baggage inspection equipment used at airports.[77] Although the detectors are relatively expensive, as costs come down they should find expanding applications. In the meantime, other techniques such as ultrasonics are being tested with encouraging results.[78] Out of all these development efforts on x-rays and ultrasonics, improved models of detectors will surely follow.

Government Regulations

Government regulations are noticeably silent on the explicit use of metal detectors. Nevertheless this equipment has been under review by the food agencies. In a proposed revision of the Current Good Manufacturing Practice, FDA in 1979 stipulated:

> Effective measures shall be taken to prevent the inclusion of metal or other extraneous material in the finished foods. Compliance with this requirement may be accomplished by using sieves, magnets, electronic metal detectors, or other suitable effective means.[79]

USDA, concerned about reports of hypodermic needles found in sausage, has studied alternative types of metal detectors. The agency is most enthusiastic about an experimental procedure which it has developed. The method measures the electrical resistance between components of meat chopping or grinding equipment. Contact between such components and any foreign metal objects reduces the resistivity, thereby indicating contamination. Such a detection system would seem to be relatively inexpensive to install. USDA hopes that manufacturers will explore this technique as a means of achieving better quality control over their products.[80]

Any food company executive who is ambivalent about metal detectors should try to imagine himself in March, 1984, as the chief executive officer of one of the manufacturers of Girl Scout cookies. In that month, reports from all over the country disclosed that pranksters had tampered with packages of these cookies. In eighty incidents across seventeen states, consumers complained of finding pins, razor blades, and other sharp objects imbedded in cookies which had been distributed by Girl Scouts in their annual drive for funds. Five relatively minor injuries were reported involving cut lips or gums.

Immediately FDA, with assistance from the FBI and state and local law enforcement agencies, began a massive investigation of the causes. FDA inspected all eight plants in the United States that produce the cookies. Each plant was reported as having had installed metal detectors to insure that no foreign objects were introduced into the cookies at the time they were sealed in their cellophane wrapping and cardboard boxes. This disclosure shifted the attention of officials to the likelihood of malevolent tampering somewhere in the distribution chain.[81] Six months later the investigation was near an end, but details were not being revealed. Whatever the causes, the forethought taken by manufacturers to install detectors and use tamper-evident packaging not only exonerated them from any suspicion but undoubtedly minimized the extent and severity of the injuries.[82,83,84]

REFERENCES

1. Aaron L. Brody, "It's in the Bag," *Chemtech*, September, 1977, pp. 559-562.
2. "Tylenol Legacy," *Time*, November 1, 1982, p. 64.
3. Susan Tifft, "Poison Madness in the Midwest," *Time*, October 11, 1982, p. 18.
4. George J. Church, "Murder by Remote Control," *Time*, October 18, 1982, pp. 16-19.
5. "Tylenol's 'Miracle' Comeback," *Time*, October 17, 1983, p. 67.
6. George J. Church, op. cit.
7. "No Plans to Require Tamper-Resistant Packaging of Food," *Food Technology*, December, 1982, p. 39.
8. Charles E. Morris, "FDA's View on Food Tampering," *Food Engineering*, March, 1983, p. 107.
9. "Copies of Revised PMO Are Made Available for Comment by FDA," *Food Chemical News*, February 7, 1977, pp. 37, 38.

10. Harold V. Semling, Jr., "Regulators Staying with Restrained Approach," *Food Processing*, January, 1984, pp. 8, 9.
11. "Heat-Shrink Seal," *Food Engineering*, December, 1983, p. 43.
12. "Tamper Evident Closures," *Food Engineering*, April, 1984, p. 60.
13. Keven Hannigan, "Baby Food Closure Grows Up," *Food Engineering*, January, 1984, pp. 97-99.
14. Ibid.
15. " 'Paper Bottles' Are Coming on Strong," *Business Week*, January 16, 1984, pp. 56, 57.
16. "Aseptic Packaging Highlights," *Food Engineering*, December, 1983, pp. 61, 62.
17. Keith A. Ito and K.E. Stevenson, "Sterilization of Packaging Materials Using Aseptic Systems," *Food Technology*, March, 1984, pp. 60-62.
18. Philip E. Nelson, "Outlook for Aseptic Bag-in-Box Packaging of Products for Remanufacture," *Food Technology*, March, 1984, pp. 72, 73.
19. Fran Labell, "Milkshake-Like Beverage in Shelf-Stable Aseptic Packaging," *Food Processing*, December, 1983, pp. 82, 83.
20. Judy Rice, "Researchers Target the Retort Pouch," *Food Processing*, March, 1984, pp. 70, 71.
21. Mark N. Vamos, "How the Food Industry Is Catering to Consumers Who Want It *Now*," *Business Week*, April 27, 1987, pp. 88, 89.
22. Judy Rice, "Answering Consumer Call for More Foods in Microwaveable Packaging," *Food Processing*, May, 1984, pp. 150, 151.
23. Judy Rice, "Squeezable Tubes & LLDPE Pouches - Efforts Intensified to Broaden Use," *Food Processing*, January, 1984, pp. 158, 159.
24. "USDA Revises Packaging Policy," *Food Technology*, March, 1984, p. 33.
25. "Tough Regs Seen for Food Package Inks," *Food Engineering*, February, 1980, pp. 30, 31.
26. "Food-Package Colors Hit Snags at FDA," *Chemical Week*, September 26, 1984, pp. 38, 40.
27. "FDA Extraction Testing . . . How and Why," *Food Engineering*, November, 1977, pp. 83, 84.
28. "Food Additive Migration Testing Evaluated," *Chemical & Engineering News*, January 23, 1984, p. 11.
29. "Potential Compromise of Package Integrity Seen with Polyterics," *Food Chemical News*, September 22, 1986, pp. 29-32.
30. Walter McQuade, "Packagers Bear Up under a Bundle of Regulations," *Fortune*, May 7, 1979, pp. 180-189.
31. "Acrylonitrile/Styrene Bottle Okayed by FDA," *Chemical & Engineering News*, September 24, 1984, p. 7.
32. "R & D/Packaging Many Changes, More to Come," *Food Processing*, April, 1984, pp. 44-48.
33. "Packaging Trends," *Food Engineering*, April, 1984, pp. 79, 80, 84.

34. Bob Hickox, "Feds Placing High Priority on Pack Regulatory Activity," *Food & Drug Packaging*, January 26, 1978, pp. 1, 14.
35. James W. Peters, "High-Strength Paper Aim - Lighter Multi-wall Bags," *Package Engineering*, August, 1982, pp. 61, 62.
36. "Anti-Slip Film Stabilizes Pallet Loads," *Food Engineering*, December, 1980, p. 51.
37. "Packaging of Grain Products," Agriculture Stabilization and Conservation Service, USDA, November 19, 1979.
38. George A. Schultz, "In-Plant Handling of Bulk Materials in Packages and Containers," *Chemical Engineering*, Deskbook Issue, October 30, 1978, pp. 29-38.
39. *Automated Bag Packaging Systems Engineered to Your Exact Needs*, sales literature from St. Regis Paper Company, West Nyack, New York, 1976.
40. "Bag's Valve, Unitizing Cut Hazardous Leakage," *Package Engineering*, June, 1982, p. 78, 79.
41. "Transportation Regulations," *Food Engineering*, December, 1978, pp. 105, 106.
42. "Proposed Performance Specification for Nonfat Dry Milk Packaging (USDA/ASCS — September, 1977)," flier from American Dry Milk Institute, Inc.
43. Charles R. Goerth, "Why Common Carriers Are Strictly Liable," *Packaging Digest*, December, 1979, pp. 25, 26.
44. Charles R. Goerth, "When Standards Go to Court," *Packaging Digest*, September, 1981.
45. *Inspection Bulletin*, IB # 79-2, Food and Drug Administration, Washington, D.C., September 28, 1979.
46. "Bin Handling of Milk Solids Allows Strict QC and Efficiency," *Food, Drug & Cosmetic Manufacturing*, February, 1981.
47. "How Land O'Lakes Moves a 'Mountain of Milk'," *Food Engineering*, June, 1983, p. 144.
48. "Palletizing, Bundling Ideas Respond to Productivity Needs," *Package Engineering*, August, 1981, pp. 27, 28.
49. "Slip Sheets: The 'Only' Way to Go," *Food Engineering*, January, 1977, p. 64.
50. Glen R. Johnson, Jr., "Slip-Sheets: Pros and Cons," *Food Engineering*, November, 1980, pp. 122, 123.
51. "Shipping Produce? Consider Slipsheets," *Food Engineering*, June, 1981, pp. 82, 83.
52. "Slipsheets Could Save $Millions," *Food Engineering*, June, 1982, pp. 70, 71.
53. "Inflatable Dunnage Bags Have Polyethylene Liner," *Food Engineering*, October, 1977, p. 151.

54. "Inflatable Bags Protect Produce While Reducing Dunnage Costs," *Package Engineering*, June, 1981, pp. 88, 89.
55. "Dunnage Bags Save on Costs, Do Job Better," *Food & Drug Packaging*, September 17, 1981, pp. 1, 59.
56. "Paper Honeycomb Prevents Shift in Transit; Takes 500 lb/sq ft." *Food Processing*, October, 1980, p. 192.
57. George A. Sim, "How to Minimize Damage in Transit," *Food Engineering*, November, 1980, pp. 124-126.
58. Cal Andres, "Weight Control Center Gives Reliable On-Line Weight Protection and Filler Control," *Food Processing*, December, 1978, pp. 84, 85.
59. Joerg W. von Beckmann and Charles R. Pettis, III, "Setting Checkweigher Setpoints," *Package Engineering*, April, 1980, pp. 69-72.
60. Robert E. Miller, "Statistics for Chemical Engineers," *Chemical Engineering*, July 23, 1984, pp. 40-44.
61. Robert E. Miller, "Distributions and Samples," *Chemical Engineering*, September 17, 1984, pp. 111-115.
62. Ibid.
63. Gerald J. Hahn, "Whys and Wherefores of Normal Distribution," *Chemtech*, August, 1976, pp. 530-532.
64. A. Kramer and B.A. Twigg, *Quality Control for the Food Industry*, 3rd ed., Volume 1, Chapter 15, Avi Publishing Company, Inc., Westport, CT, 1970.
65. Frederick R. Morin, "Finding the Right Target Weight Can Minimize Product Giveaway," *Food Engineering*, August, 1978, p. 152.
66. Morton E. Bader, "Quality Assurance — III Using Statistics," *Chemical Engineering*, June 16, 1980, pp. 123-129.
67. C. Kloos and W.L. Clark, "Statistical Provisions of Federal Net Content Control Programs," *Food Technology*, October, 1981, pp. 50-57.
68. *Checking the Net Contents of Packaged Goods*, Handbook 133, National Bureau of Standards, Washington, D.C., 1981.
69. *The Principles of Checkweighing*, Hi-Speed Checkweigher Co., Inc., Ithaca, New York, 1976.
70. "Supreme Court Decision Due on Short Weight Packaging," *Food Engineering*, May, 1976, p. 24.
71. "Weights & Measures: How State Laws Vary," *Food Engineering*, June, 1978, pp. 145, 146.
72. "States Can't Set Weight-Labeling Rules Stiffer Than Federal Laws, Justices Say," *The Wall Street Journal*, March 30, 1977.
73. "New Checkweigher Is Exceptionally Easy to Set up and Operate," *Food Processing*, October, 1976, p. 140.
74. "Checkweigher Fine-Tuning," *Package Engineering*, November, 1978, pp. 50-52.
75. "Detects Metal Particles As Small As .03″ at Speeds from 9 to 600 fpm," *Food, Drug & Cosmetic Manufacturing*, April, 1978.

76. *Look Inside!*, sales brochure from Scanray Corporation, Harbor City, CA, 1984.
77. "X-ray Sorter Detects and Rejects Rocks and Foreign Matter in Almonds," *Food Engineering*, October, 1983, p. 131.
78. "On-Line Ultrasonic Monitor Detects Contaminants in Fluids," *Food Engineering*, July, 1979, p. 104.
79. *Federal Register*, June 8, 1979, p. 33247.
80. "New Metal Detection Method Developed by USDA Eyed by Processors," *Food Chemical News*, January 10, 1983, pp. 20, 21.
81. Susan F. Rasky, "X-Rays Aid Drive on Tainted Girl Scout Cookies," *The New York Times*, March 31, 1984, p. 6.
82. "Inquiry on Cookie Tampering by U.S. Reported Near End," *The New York Times*, August 26, 1984.
83. "Tainted Scout Cookies Spur Indiana Warning," *The New York Times*, March 15, 1986, p. 9.
84. Felix Kessler, "Tremors from the Tylenol Scare Hit Food Companies," *Fortune*, March 31, 1986, pp. 59, 62.

CHAPTER TWELVE
WAREHOUSING, SHIPPING, AND FOOD SERVICE

The growth in mass marketing of food is reshaping distribution with the result that warehousing, shipping, and food service are becoming more interconnected. Today such giant supermarket chains as Safeway Stores operate their own warehouses and shipping facilities. Computerized information systems provide Safeway's management with up-to-the-minute reports on the availability of perishables, as well as on product sales, delivery schedules, truck routing, and the allocation of shelf and storage space in its 2,507 stores and supporting distribution centers.[1] Fast food franchisers are epitomized by McDonald's Corporation, which in 1983 served its 45 billionth hamburger to some grateful customer in one of its 7,778 worldwide restaurants. This success is largely attributed to the company's motto, Q.S.C. & V., which stands for "*Quality* food products; efficient, friendly *Service*; and restaurants renowned for the *Cleanliness* and *Value* they provide."[2] Even the family-run delicatessen, which stubbornly clings to its independence, is relying more than ever on integrated wholesalers for its needs.[3,4] As the mass marketing trend continues, management is provided new opportunities to institute controls over such vital phases of its operations as quality assurance.

WAREHOUSING

The management of food warehouses requires effective controls over product that is inbound, outbound, and in storage. These controls encompass business activities such as billings, inventory reporting, and the assessment of warehousing fees. Controls must also be exercised over factors related to quality assurance, including breakage, contamination, and stock rotation. In practice there are close ties between many of the controls used for accounting purposes and those for quality considerations so that a distinction between them is not always practical nor necessarily desirable.

Inbound Shipments

Inbound shipments must be carefully inspected for damage, evidence of insect and rodent activity, filth and debris, and contamination by any harmful substances. The procedure for handling adjustments was outlined in Chapter Ten under the investigation of service complaints. Any irregularity should be brought to the attention of the carrier. In the case of a delivery by motor carrier, the driver should be asked to corroborate any loss or damage by a signed statement on the Delivery Receipt. For rail shipments, the carrier should be notified and requested to make an inspection. Any disagreement or exception should be duly noted on the Railroad Inspection Report, which is submitted by the carrier on completion of its investigation. The warehouse should make an accurate count of all goods received, regardless of their condition, and it should reconcile these numbers with each billing. When inspecting incoming shipments the Food and Drug Administration (FDA) offers these tips:

- Note outside condition of carrier.
- If doors of vehicle compartments have a seal, note if it is broken.
- Break seal and open doors — at the same time note odor and temperature.
- Note condition of stacked cartons or other containers.
- Look for evidence of insect, rodent and bird activity.
- Remove random samples of food containers for product examination.
- While unloading, note if non-food items are also in the shipment.
- After unloading, observe inside condition of carrier.[5]

Outbound Shipments

Outbound shipments require just as much attention as incoming receipts. All cars and trailers must be thoroughly inspected as will be discussed later in this chapter under the section on shipping. An inspection report should be placed on file with a copy of the bill of lading. The warehouse should carefully examine the condition of all containers and hold back any that are damaged or in poor condition. Containers that are taped to cover up punctures or leaks must not be shipped. Before non-conforming product may be shipped, written instructions must be received to this effect. A detailed record should be kept of the quantities of each product shipped and their code dates (lot numbers). Failure to record the code dates will only make the job of tracing shipments, in the event of a product recall, that much harder. All doors, as well as loading and unloading outlets, should be secured with seals, and the seal numbers should be recorded on the bill of lading.

Product in Storage

Product in storage must be held under sanitary conditions and away from all toxic and harmful substances. To remind shippers and warehouses of the need to follow this precaution, certain food manufacturers have adopted the use of distinctive warning labels and stickers as shown in Figure 12.1. Printed on fluorescent orange paper stock, these notices display a unique symbol, consisting of a triangle enclosing a bull's-eye, to catch people's attention. Such labels are to be affixed to all pallets of food products or to loose containers loaded on a truck or car. The labels should be conspicuously placed so that they are immediately visible when the door of the van or car is opened. The corresponding stickers are designed for attaching to copies of the bill of lading directed to the attention of the carrier and the consignee. Neither warning labels nor stickers should be placed so as to obscure important information on the containers or documents.

NOTICE—DO NOT CONTAMINATE
DO NOT SHIP OR STORE WITH POISONOUS OR ODOR
EMITTING ARTICLES—THIS PRODUCT IS INTENDED
FOR USE AS AN INGREDIENT IN FOODS OR FEEDS.

NOTICE

DO NOT SHIP OR STORE WITH POISONOUS OR ODOR-EMITTING ARTICLES

This product is intended
for use as an ingredient in
foods or feeds.

DO NOT CONTAMINATE

FIGURE 12.1.
FOOD CONTAMINATION WARNING STICKER AND LABEL

Management control is easier to maintain over company owned or operated warehouses than public ones. Company warehouses may be economical where volume is substantial and predictable. Procedures are available to determine the strategic locations for such facilities.[6,7] In many circumstances, however, the additional capital investment or the required operating expenses for a company warehouse cannot be justified. The only recourse then is to find a public warehouse that meets all the requirements and standards for a food facility. A contract should be drawn up with such a warehouse, stating all the terms and conditions, including a promise to comply with FDA regulations.

Voluntary Industry Sanitation Guidelines

Food regulations covering warehouses are one and the same as Current Good Manufacturing Practice provisions (21CFR110). These have been discussed in Chapter Five. Of course, only those sections which are applicable to warehouses would be of concern to distributors. The principal points have been reviewed in an excellent publication, *Voluntary Industry Sanitation Guidelines for Food Distribution Centers and Warehouses.*[8] This brochure was prepared in 1974 by leading food associations in cooperation with FDA. It suggests programs that can be instituted by management, and it highlights the topics that should be covered in regular warehouse inspections. Realizing that each situation has its own special needs, the publication urges each operator to develop his own set of guidelines. The most important rules for warehousemen to remember were probably best summarized by FDA in a 1966 poster:

1. Promote personal cleanliness among employees.
2. Provide proper toilet and hand-washing facilities.
3. Adopt "good housekeeping" practices.
4. Keep food handling equipment clean.
5. Reject all incoming contaminated foods.
6. Maintain proper storage temperature.
7. Store foods away from walls.
8. Rotate stock and destroy spoiled foods.
9. Do not use or store poisonous chemicals near foods.
10. Maintain an effective pest control program.[9]

Stock Rotation

Of the ten rules for food warehousemen listed above, none is more important than the requirement to rotate stock. Using accounting vernacular, product should be shipped on a First In First Out (FIFO) basis. This principle has been referred to as a heuristic program, one which is difficult to prove, but it consistently gives good results.[10] In any event, product is not to be shipped Last In First Out (LIFO), and if product is "FISH" (First In Still Here), the warehouse is in real trouble.

Without support from the bookkeeping department, warehouse management will have difficulty in being absolutely certain that stock is being properly rotated. A smaller warehouse might attempt to keep track of product by staging inventory on the floor in the order in which it was received. This method, however, is far from infallible. Effective inventory management also requires that all damaged or unacceptable product be immediately removed to a salvage area and tagged as not shippable. Records should be kept current as to the status of all product in inventory. In spite of their physical restrictions and considerable investment, automated warehouses are a trend of the future. These facilities as they become available should be of assistance in maintaining proper rotation of stock.[11]

Notwithstanding the adoption of all the above safeguards, product can be misplaced, overlooked, or forgotten in even the best managed warehouse. A few visits to warehouses are all that is required to confirm this unfortunate state of affairs. Therefore, borrowing a concept from the Japanese called "zero" or "just-in-time" inventory control[12], the following procedure is suggested for warehouse management:

1. A level of inventory should normally be maintained sufficient to handle customers' demands but not in excess of these needs. To assist the manager responsible for inventory control, printouts should include tabulations not only of inventory levels but also the current turnover rate or sales figures. From these data the months of supply on hand can be calculated.

2. In determining the right level of inventory to carry, one must consider the shelf life of the product, seasonalness of sales, and delivery time to restock inventory. In addition, the following external factors should be taken into account: the possibility of labor strife, price adjustments, shortages, and any other anticipated upsets.

3. Reported inventory can be overstated in terms of quantity or value, and this possibility should be recognized. Stock may be old, deteriorated, obsolete, damaged, out of specification , or nonexistent. These conditions can have an impact on quality control and on finance.

4. Unreported inventory may actually be present in the warehouse. These goods can cause quality control problems, but they do not constitute a financial burden or risk. Such inventory which is not on the books will not be subject to taxes, storage fees, carrying charges, and most important, it does not pose a threat of a writedown.

5. In order to determine the quality of reported inventory and to assure stock rotation, on a periodic basis, the inventory on the books should be temporarily depleted to zero. This action should be coordinated with the marketing department to guarantee uninterrupted service, e.g., by scheduling during a lull in demand or by arranging an alternate source of supply. Without taking an expensive physical inventory, this method can ferret out any overvalued product, assuming the warehouse has instituted proper controls over outbound shipments. As confirmation, a cursory inspection of the warehouse can be made after the reported inventory has been dropped to zero and before product is restocked.

SHELF LIFE

In the discussion of food preservation in Chapter Two, various techniques were reviewed for extending the shelf life of foods by retarding spoilage. These concepts were expanded in the last chapter on packaging, which included material on aseptic packs and retort pouches. In addition, open date labeling was discussed in Chapter Four as a means of conveying information to consumers about the shelf life of a perishable food product. Notwithstanding these prior references, shelf life is so important to food distribution that the subject requires some unifying comments.

Limiting Factors

Shelf life can be limited by a number of different factors. Spoilage due to microbial growth or enzymatic reactions is probably most common. Aside from these causes, many food ingredients degrade over a period of time. The reaction rates of some leavening acids change with time, and the viscosity of certain stabilizers, such as carrageenan, may be reduced with storage. Vitamin A and riboflavin lose their potency when exposed to light, while the artificial sweetener, aspartame, gradually decomposes in carbonated beverages. The nutritional quality of proteins has been found to fall the longer these substances are held and the higher the temperature and humidity.[13,14] Physical changes can also limit the shelf life of products. Powdered materials can become caked so hard during storage that, for all practical purposes, they are unusable.

In every case where shelf life is limited, packaging and storage conditions have a profound effect. Higher temperatures and humidity invariably are harmful. Thus, the caking tendency of certain powdered products is accentuated when they are stored under tropical conditions. If these products are packaged in multiwall bags and stacked several pallets high, the bags on the bottom most likely will solidify. There are indications too that unstable conditions or the cycling of such a critical variable as temperature can be detrimental to some products by accelerating physical changes.

Because storage conditions have a direct bearing on shelf life, instructions for proper storage should be clearly marked on containers and included with product specifications. For example, protein products and hygroscopic materials should be labeled, "Store in cool, dry place." Such a directive does not necessarily require expensive air conditioning in warehouses although the common practice of installing climate control in supermarkets for the comfort of their patrons most surely benefits the keeping properties of many food items. Vitamin blends containing Vitamin A or riboflavin need the warning, "Protect from exposure to light." Some products, e.g., sodium erythorbate, which may slowly degenerate in the presence of air, should carry the notice "Replace cover after use." Perishables, whether USDA inspected or not (cf. Chapter Four), should be labeled "Keep refrigerated" or "Keep frozen" as need be.

Shelf-Life Studies

Well designed studies are required to determine the shelf life of a product. To speed up results, accelerated shelf-life testing procedures can be employed.[15] For example, compaction tests, in which high loads are applied to powder materials, may be used to assess their proclivity to form lumps. To test other properties, product may be stored under "jungle room" conditions (usually 90 °F and 75 percent relative humidity) in order to obtain quicker results than would be obtained during actual storage. When field testing is desired, these experiments may be conducted in southern states such as Florida where extremes in temperature and humidity are experienced.

During shelf-life studies, the product is periodically sampled and analyzed for its limiting attribute. To illustrate, the critical property could be the potency of an added vitamin. The data so obtained will yield a curve of potency versus storage time. Then for any arbitrarily selected shelf life, the amount of degradation of the vitamin can be determined. To make up for this quantity lost during storage and distribution, an excess quantity of the ingredient must be added initially to the food product. By increasing the level of fortification, the product will stay within specification up to the time of its established shelf life.

Unlike some attributes, shelf life is not an additive property. In Chapter Two certain properties, such as nutrient level or percentage of impurities, were found to be additive in mixtures. By contrast, the shelf life of a blend is determined by the least stable ingredient. Thus, the shelf life of a cake mix containing sugar, flour, vegetable shortening, leavening, emulsifier, salt, and artificial coloring would equal that for the flour. The shortening, which is partially hydrogenated, probably has good oxidative stability, and the other ingredients are even less sensitive to deterioration.

Shelf-Life Specifications

The judicious choice of terminology for shelf life can avoid confusion. If it is expressed in absolute terms, such as 6 months or 1 year, a question may arise concerning the proper course of action to take after the stated time has elapsed. Should the product be returned, destroyed, or used anyway? This dilemma can be avoided by stating the shelf life as "in excess of" or "over" a given length of time. Because manufacturers have an interest in specifying shelf life conservatively, a product may be perfectly wholesome and safe to use for a longer period of time, particularly if storage conditions are ideal. By the same reasoning, foods like salt that will keep indefinitely probably should not be assigned an indefinite shelf life but rather some fixed value, e.g., "over two years."

Even though shelf-life data may be expressed in terms of ranges or limits, they are extremely useful guidelines. They underline the need to keep product moving efficiently through the distribution pipeline. The relationship between shelf life and storage times is given by the following expression:

$$T_{SL} \geq t_M + t_D + t_C$$

where T_{SL} = shelf life
t_M = time in manufacturer's plant
t_D = time in distribution
t_C = time in customer's hands

For any product with a short shelf life, delay should be minimized at each step in the handling of the product. Obviously, if the months of supply of a product in a distributor's warehouse exceeds ($T_{SL} - t_M - t_C$), some of the product will be older than its shelf life at the time of use.

Controlled Atmosphere Storage

Controlled atmosphere (CA) conditions have been used to extend the shelf life of perishable commodities. This result is accomplished by replacing the air surrounding the product with various gas mixtures containing carbon dioxide, oxygen, nitrogen, carbon monoxide, ethylene, propylene or acetylene.[16] A variation of this approach is the use of an hypobaric environment otherwise referred to as vacuum packing.[17] CA has several salutary effects. It is capable of killing insects as in grain fumigation[18], inhibiting bacterial growth[19], and extending the freshness of respiring fruits and vegetables.[20,21] Experiments have shown that this modification can reduce the spoilage of milk by psychrotrophic bacteria.[22] A novel way of obtaining a CA environment is to package produce in semi-permeable films through which gases diffuse at different rates.[23]

UNIVERSAL PRODUCT CODE

Product identification (ID) numbers are necessary to improve the efficiency of distribution; these numbers will help to avoid costly mistakes in product identity. If product is not clearly marked, a warehouse operator may have trouble in picking the right containers from among similar products. Often the only difference between products is the coarseness of the material or the characterizing flavor, and information of this sort is usually difficult to spot on a label. Besides performing a vital business function, the ID number is required in quality assurance. Together with the code date (lot designation), this number serves as the product's "fingerprint" on all quality assurance records.

Color Coding

Various approaches have been tried for product recognition. One of the simplest is color coding, which is extremely effective, for example, in differentiating between grades. By using different colors of ink — red, blue, black, gold — for printing labels, quick identification is possible. Unfortunately the number of colors in the rainbow has a practical limit, and lighting conditions in warehouses are not always perfect for recognizing different hues. In spite of these restrictions, color coding, when used in conjunction with other systems, can be very helpful.

Assignment of Product Numbers

The next degree of sophistication in product identification is the assignment of numbers to products. At first, this approach might seem simple, but on further reflection the problems are considerable. The following indexing procedure emphasizes important points to consider.

1. Each product ID number must be unique. One item cannot be given two codes. And vice versa, one code must not be assigned to two or more products.[24]

2. The format of the number should be direct and easy to recognize. Thus, it should be as short as possible without limiting its usefulness. Five digits have been found to be a good compromise. To avoid possible confusion with alphanumerical code dates, only numerals should be used in its makeup.

3. A workable plan for assigning numbers should be developed. Large business often have many numbering systems for different purposes. For maximum flexibility and to avoid confusion, product ID numbers should correspond to codes used in the organization's accounting practice. In some cases this plan may require truncating a standardized code to arrive at a five digit number.

4. An ID number should be clearly marked as such for quick reference. The prefix "Product Number," "ID No," or even just "#" will help to guide an interested party.

5. All labels and packaging should display the ID number conspicuously but not in a position that will interfere with information required by law to be on the label. Convenient locations for ID numbers are in the upper corners of labels and on the gussets or butt ends of bags.

6. Records, reports, shipping documents and product literature should include the ID number whenever it can be of assistance.

Bar Codes

As useful as ID numbers are, they have one serious drawback: they cannot be read by automated equipment. To correct this shortcoming, a third generation system has been developed. Now a familiar household sight, the Universal Product Code (UPC) is a cryptic-looking bar symbol that can be scanned at the checkout counters of supermarkets. (Figure 12.2) It encodes a number containing ten digits, five of which identify the manufacturer and five the product. This ingenious creation is compact, requiring a minimum of space on the label, and it is forgiving of minor imperfections in its rendition.[25]

After a slow start, UPC is sweeping the food industry. By 1984 close to one-third of the 30,000 supermarkets in the United States were equipped with automatic scanners.[26] With adaptations to handle the random-weight packages of fruits, meats, and custom sliced cheese wedges, virtually all supermarket items can at

Manuf.   Product
Number   Number

FIGURE 12.2. UNIVERSAL PRODUCT CODE

present be accommodated.[27] Although the initial impetus for installing UPC scanners was to improve service and reduce labor costs, its potential is much greater. UPC provides the means by which on-line control can be established over the entire distribution function, from purchasing down through warehousing, shipping, and retailing.[28]

SHIPPING

A boxcar was termed a "warehouse on wheels" by FDA Commissioner Alexander M. Schmidt at a 1974 conference on railroad car sanitation.[29] This parallel was drawn to point out the necessity of protecting food from contamination regardless whether it is stationary or in transit. Efforts, however, to correct the widespread unsanitary conditions found among haulers have been frustrated by the shared responsibilities within the private sector and by overlapping jurisdictions between government agencies. Four years later in 1978, with little progress in sight, a plea was made at another meeting on carrier sanitation that shippers, carriers, receivers, and government regulators stop passing the buck.[30]

Dismal Record

The dismal record of car sanitation has been well documented. In an incomplete 1977 survey, FDA reported that 49 rail cars had either minor or gross defects out of 731 cars inspected. In the same study, 34 trucks from a total of 828 were found to be in unsatisfactory condition.[31] A year later the Interstate Commerce Commission (ICC) disclosed that 98 rail cars in a survey of 422 cars in food service had not been completely unloaded by the consignee and contained debris or scraps of food. (Commission rules require that a railroad is not to release a car from demurrage until it is completely unloaded.) The statistics obtained by ICC on trailers showed that 112 out of 465 were incompletely emptied.[32] Supporting the general findings of FDA and ICC, the Association of Operative Millers reported that, in a 1978 survey, close to 40 percent of the boxcars offered by railroads to food shippers had defects compared with 39 percent in 1972.[33]

Enforcement of rail car and trailer sanitation has been hampered by legal hair splitting. In spite of a memorandum of understanding between FDA and ICC to report suspected violations on a reciprocal basis, FDA has intimated that ICC alone "has some unique strengths which can be used to resolve" these problems.[34] In 1980 the outgoing chief counsel at FDA opined that the agency does not have any course of action against a carrier that delivers an unsuitable empty car unless the shipper can be expected to accept it for loading food articles. If the shipper always rejects an unfit conveyance, no food adulteration is likely to occur and the carrier has not contributed to a prohibited act.[35] Even though FDA does not have a free hand in every instance, the carriers' immunity from prosecution has been significantly restricted by a key court decision.

In a case going back to 1974, the Penn Central Railroad diverted an uncleaned car from its prior use, namely, moving non-food materials, to the shipping of corn gluten meal. CPC International, the shipper, did not notice anything wrong with the car and thereupon loaded it with corn gluten. Subsequently the corn gluten was found to be contaminated with lead oxide, which was traced back to the uncleaned car. Even though the prosecution charged that CPC International was legally at fault for introducing an adulterated food into interstate commerce, the court found the railroad guilty instead. This verdict means that carriers now can be held accountable for the contamination of food products during shipping.[36]

Voluntary Transportation Guidelines

Realizing that legal enforcement of sanitation is slow and inefficient, FDA has attempted to marshal support for a voluntary compliance program. For this purpose the agency organized the Rail Car Sanitation Action Committee (later changed to Carrier Action Committee) which consisted of representatives from FDA, ICC, USDA, Department of Transportation (DOT), the railroads, and the rail-service users.[37,38] As an outgrowth of the effort, the Association of Food and Drug Officials proposed standards for Good Transportation Practice (GTP). These proposals were later incorporated into the 1976 *Voluntary Transportation Guidelines* designed to prevent the contamination of food, feed, and related products.[39] Endorsed by eight industry and regulatory associations, and issued in cooperation with FDA, these guidelines offered the following definitions:

1. A *clean car* is (a) free from evidence of vermin infestation (including but not limited to birds, rodents, and insects); and (b) free from debris, filth, visible mold, undesirable odors, and evidence of residues of toxic chemicals.
2. A car in *good repair* should have sound interiors and exteriors, including doors and hatches that are tight-fitting and, when closed and sealed, capable of excluding rodents, birds, and other pests.[40]

The most significant feature of the *Voluntary Transportation Guidelines* (Figure 12.3) is that they attempt to pinpoint responsibilities among shippers, carriers, and receivers. The guidelines confess to some redundancy in the procedures but emphasize that the double-checking is important in order to maintain the high standards that are required. While the guidelines avoid a discussion of rail car specifications or maintenance requirements, they do endorse the maximum use of Exclusively-for-Food (XF) cars. The interiors of these rail cars are lined with a white epoxy coating that is easy to clean. XF cars are to be dedicated to the transportation of "processed, packaged food and food products only, except that other non-contaminating commodities enclosed in clean packaging" may be loaded in the same car.[41]

FIGURE 12.3.

VOLUNTARY TRANSPORTATION GUIDELINES *

These guidelines contain recommendations for shippers, rail carriers, and receivers concerned with the prevention of contamination of food and feed. They are intended to apply specifically to boxcars and covered hopper cars which are used for transporting such products.

I. CAR ORDER
    (Shipper Responsibility)
    A. Place car orders with appropriate railroad personnel.
    B. Specify in each order:
       (1) type and size car required (e.g. Class A-50' boxcar, airslide car, etc.);
       (2) the commodity to be loaded;
       (3) whether commodity is bulk or packaged;
       (4) date required;
       (5) location (track and door number if applicable) where car is to be spotted; and
       (6) load destination and route (if known).

II. CAR FURNISHING
    (Carrier Responsibility, except as noted)
    A. Where cars are required for transportation of food, feed or related items, furnish cars suitable for the intended purpose and provide cars which are in a clean condition, in good repair, and of adequate design and construction for the intended purpose.
    B. In furnishing free-running cars, take necessary precautions to insure that the car is suitable for the intended purpose.
    C. Where cars are dedicated to food, feed or related product category use, furnish cars with doors and hatches closed and sealed.

(continued)

(Shipper Responsibility)

D. Where a car is assigned to a shipper for its exclusive use, the shipper has the responsibility of inspecting and maintaining such car in a clean and sanitary condition.

III. CAR LOADING
(Shipper Responsibility)

A. Inspect all rail cars offered or intended for loading to determine if they are clean and in good repair. Refrain from loading any rail car deemed unsuitable until such time as all noted defects which may contribute to contamination are corrected. Such defects may include:
   (1) damage to floors, walls, ceilings, doors, and hatches;
   (2) protruding nails or bolts;
   (3) dunnage, trash, or other debris;
   (4) residue of prior loading;
   (5) evidence of contamination by prior toxic material loading;
   (6) vermin infestation or visible mold; or
   (7) objectionable odors.
B. Whenever defects noted in a rail car are not corrected at the shipper's plant, reject the defective rail car to the rail carrier, stating the reasons for rejection; maintain inspection records of all defects causing rail cars to be rejected.
C. Load only products which are themselves uncontaminated and are free from substances or components which are likely to contribute to contamination of other products in the load during transit, or which are likely to result in contamination of the rail car.
D. All packaged food and feed products are to be appropriately packaged and loaded in order to minimize physical damage or contamination under reasonable transportation conditions and procedures.
E. In the loading of products, take adequate precautions to minimize contamination of the rail car through hatches, pipes, hoses, vents, conveyors or other potential routes of contamination.
F. See that persons responsible for the loading operation take all other precautions, as may be appropriate, to protect the integrity of both the transportation equipment and its contents.
G. Close and seal all doors and hatches and tender billing instructions to carrier.

IV. CAR TRANSPORTING AND DELIVERY
(Carrier Responsibility)

A. Remove car from shipper's siding and transport cars to destination with all due care for the integrity of the lading.
B. Use reasonable diligence to prevent unauthorized entry into cars. Maintain all seals intact and all doors and/or hatches secured.
C. In the event of derailment or other type of major accident or damage, natural catastrophe (such as a flood), or detection of unauthorized entry into car, promptly notify shipper and receiver.
D. Notify receiver that car has arrived and spot car as per receiver's instructions.
E. If a car is in shipper's assigned service or is dedicated to food, feed or related product category-use, notify receiver of this fact.

V. CAR UNLOADING
(Receiver Responsibility)

A. Examine all incoming rail cars carefully to determine if doors, hatches, and seals are intact and untampered with upon arrival at the delivery point.
B. Record the seal numbers of the doors and hatches prior to their removal. Note any broken or damaged seals and report such findings to the rail carrier and shipper.

(continued)

FIGURE 12.3. (*continued*)

C. Upon opening and prior to unloading of the product, examine the exposed interior of the rail car for evidence of any detectable signs of potential contaminants and adulterants including but not limited to insects, rodents, mold or undesirable odors. Continue this examination during the entire unloading operation.

D. In the event contaminants and/or adulterants are noted:
   (1) Notify rail carrier to make an inspection and provide an inspection report.
   (2) Notify shipper for disposition.
   (3) Where the shipment contains damage or contamination which could lead to contamination of the receiver's establishment, do not permit the product to enter the building; in other cases, separate damaged or contaminated product from the remainder of the load.
   (4) Keep a record indicating the type and disposition of damaged, adulterated, and deteriorated product, and of rail car.

E. Prior to releasing a car to the carrier as an "empty," completely remove all products, including damaged or refused product, dunnage, debris, and other materials connected with the inbound shipment from the rail car.

F. Prior to releasing car to the carrier, report all contamination, physical damage, or other conditions incompatible with further use of the car for food, feed, and related products to the carrier.

G. Replace and/or secure all bulkheads and other appurtenances that are a part of the rail car prior to releasing the car to the carrier.

H. Close all doors and hatches. When carrier's agent notifies you that a car is in shipper's assigned service or is dedicated to food, feed or related product service, seal the car after complete unloading.

VI. REMOVAL OF EMPTY CAR AND SUBSEQUENT HANDLING *(Carrier Responsibility)*

A. Ascertain that the car has been completely unloaded and emptied as as required. A car sealed by the receiver is considered to have been completely unloaded and emptied as provided in Section V(E).

B. Do not remove car if car is not completely unloaded and free from product, dunnage, or other debris.

C. Upon notification by receiver that a car contains contamination or physical damage, take necessary action for cleaning and/or upgrading the car before returning it to food, feed or related product category-use.

The *Voluntary Transportation Guidelines* sidestep a number of specifics, such as what constitutes "appropriately packaged and loaded." To fill the silence, some receivers have issued detailed instructions which they ask their suppliers to follow. For example, shippers might be expected to line the floor, walls, and ends of rail cars and trailers with kraft paper or corrugated fiberboard whenever (a) bags are deadloaded (loaded individually), (b) this practice is requested by a customer, or (c) the condition of the interior would be acceptable after such lining. Staples, tacks, or nails should never be used in fastening the paper. By convention, when shipping via truck, only closed vans should be used. Flat-bed trucks, tarpaulin-covered trailers, or other unprotected vehicles should not be substituted. If a customer insists that a shipment be made in one of his trucks which fails to comply with GTP, the shipper should obtain a signed release from the customer absolving the shipper of all responsibility.

Bulk Shipments

In bulk shipments, the rail car or trailer essentially serves the dual purposes of packaging and conveyance. Thus, all safeguards that would be observed for inspecting containers should be applied to hopper cars or trucks, tank cars and tank wagons. As required, these units should be cleaned by steaming, washing with detergent, rinsing, and drying. When a car is transferred to food use, it may have to be sandblasted clean.[42] Although rubber-lined tank cars are cheaper and may be adequate for limited usage, as a rule stainless steel equipment provides better service for food products.

In checking cars and trailers used for bulk shipments, the shipper should make sure that all appurtenances are clean and in good working condition. Valves should be operative and non-leaking. Hatch covers and gaskets must be free of dirt, undamaged, and capable of sealing properly. Valves, unloading pipes, and ports shall be purged and tightly closed before loading product. All exterior bulk handling openings shall be clean, tightly sealed and capped to protect them from recontamination during transit. Unloading hoses must be sanitary and covered at both ends when not in use. If a car contains a steam coil for remelting product, it should be in good repair. To summarize, all lines and fittings attached to the vehicle need to be maintained in A-one condition.

Mail Orders

Shipping by mail is a special case which requires some comment. Food delicacies ranging from fresh fruit to shelled pecans are being distributed in the "direct response market."[43] The growth in food-by-mail sales has been nothing less than stupendous. This acceptance has been achieved by gaining the trust of consumers. The greatest concern of the business is to keep the level of complaints low, especially those related to the arrival of food in poor condition. To head off such complaints, suppliers spend a disproportionate sum on packaging. One distributor of filet mignon steaks packs these cuts in dry ice and places them in an insulated cooler that can be reused by the customer for parties, camping, and tailgate picnics.[44] Such premium packaging is the best guaranty for holding on to satisfied customers.

FOOD SERVICE

In this discussion, the scope of food service includes not only restaurants, lunch counters, and institutional facilities, but also retail food stores, vending machines and other food outlets catering to the public. As the eating habits of Americans change, the demarcations between these establishments are becoming blurred. Even so, by 1980, there were reported an estimated 165,000 food stores and 350,000 restaurants in the United States. On an average day these food establishments served 150 million meals outside of the home. With all this activity, the statistics for foodborne illness caused by the mishandling of food should not be surprising. The Center for Disease Control disclosed in 1977 that such outbreaks were attributed 73 percent of the time to food service establishments, 25 percent to home kitchens, and 2 percent to food manufacturers.[45,46]

Local Health Boards

Traditionally, the surveillance of food service establishments has been in the hands of local and state government agencies. The City of Baltimore is credited with establishing the first health board in the United States around 1798. Since that time the number of autonomous food regulatory bodies has increased in number to about 3,500. Not until 1934 did the federal government begin to take an active interest in food service standards. Working in an advisory capacity, the U.S. Public Health Service and then FDA became more involved in setting guidelines for the industry. This effort resulted in the publication of three pivotal FDA documents: *The Vending of Food and Beverages* (1965), *Food Service Sanitation Manual* (1976), and *Retail Food Store Sanitation Code* (1982).

Model Codes

The three publications on vending machines, restaurants, and food stores each incorporate a code or model ordinance for adoption by the states and local health boards. Even though the *Food Service Sanitation Manual* did not get into print until 1978, 37 states and 1,100 county and municipal regulatory agencies had enacted its provisions by mid-1982. In due time, the *Retail Food Store Sanitation Code* is expected to have a similar impact.[47] The three model codes replaced earlier FDA proposals to issue Good Manufacturing Practice (GMP) regulations covering the same fields.[48]

In 1986 FDA proposed developing a model Unicode which would update and combine the existing model codes on food service, stores and vending. It would also extend to other areas of food retailing, including mobil and temporary food service as well as bulk self-service. FDA expected that the new Unicode would eliminate inconsistencies in the present codes and reduce duplication and administrative costs. Provisions in the new code would reflect performance criteria as much as possible instead of specification standards.[49]

The FDA sanitation codes in essence elaborate on the principles laid down in the umbrella GMP. Of special interest, however, are the required temperatures for holding and preparing food. Potentially hazardous foods requiring refrigeration shall be kept at 45 °F (7 °C) or below, and frozen foods must be maintained at 0 °F (−18 °C) or below. When served hot, potentially hazardous foods must be held at 140 °F (60 °C) or above. In cooking potentially hazardous foods, all parts must be heated to at least 140 °F, except poultry products shall be heated to 165 °F (74 °C) or more, and pork to 160 °F (71 °C). Rare roast beef shall be cooked to an internal temperature of at least 130F (54 °C). Reheating potentially hazardous foods that have already been cooked and subsequently refrigerated must be done rapidly, bringing them to 165 °F or higher before serving. Sensitive frozen foods may not be thawed by allowing them to stand at room temperature but must be handled by prescribed methods.[50,51,52,53]

Compliance Provisions

Compliance provisions are important aspects of the sanitation codes. One of the most effective means of controlling establishments is by requiring operating permits or licenses, which may be suspended or revoked for good cause without delay. Furthermore, plans for new facilities or modifications of existing structures must be submitted to the proper regulatory authorities and approved before work on them can begin. These plans must provide detailed information about all design features affecting sanitation. Finally, all food service establishments shall be open for inspection by accredited personnel at any reasonable time. Food may be examined and sampled to determine compliance with the codes. Taken together, these compliance procedures provide tight supervision of all establishments.

Employee Training

In the area of employee training, the FDA sanitation codes are proving useful. The Food Marketing Institute (FMI) has incorporated this material into its training program for food store managers, and the National Institute for the Foodservice Industry (NIFI) is offering a similar course for restaurant employees.[54] The need for this kind of education was emphasized in a paper given at the 1984 Annual Meeting of the Institute of Food Technologists (IFT). This presentation pointed out that food service employees are woefully inexperienced because of their low average age and the high turnover rate.[55] Recognizing this problem,

McDonald's Corporation has built an ultra modern campus for its Hamburger University, which by 1977 had matriculated more than 10,000 students. FDA wants to see the training of food service managers upgraded, and against some industry resistance the agency has proposed a nationwide certification system for these professionals.[56]

All of the missionary efforts in employee training are beginning to show some results. The first round of testing in the Food Protection Certification Program to certify people who supervise retail food operations commenced in August, 1985. This test is administered by the Center for Occupational and Professional Associations, an arm of the Educational Testing Service in Princeton, New Jersey. The test is voluntary, but those who take it and pass will be listed in a national registry. The program has the support of industry, academic and regulatory officials.[57]

Patchwork of Regulations

In spite of the steps taken to raise the standards of the food service industry and to rationalize its operations, general dissatisfaction with the progress has been expressed. Several fast food chains are chafing at the want of uniformity in the sanitation codes from one jurisdiction to another. Other complaints are leveled at the lack of even-handedness and objectivity on the part of some local health boards. Restaurant chains, that are seeking economies of scale, find the differences in design codes particularly vexatious. The added expenses and delays associated with expansion plans have been particularly bothersome to chain operators.[58,59,60]

The answer to the checkered regulation of the food service industry has proven to be elusive. There are loud voices calling for the preemption of state and local ordinances by federal statutes. A more considered viewpoint was expressed by Chambers F. Bryson, Chief, Food and Drug Branch, California Department of Health Services. He advocates a "planned cooperative program" involving the three levels of government. The program envisions that states and local agencies would enter into Memorandums of Understanding (MOU) that will define the responsibilities of each contracting party. On a higher plane, FDA would implement plans, including the execution of MOU's, to assure consumer protection in the most efficacious manner possible.[61]

Salad Bars

Some type of planned cooperative program between federal, state, and local agencies is necessary not only to solve existing problems but to tackle new issues as they arise. One such concern is over the rapid growth in the number of salad bars and the proper sanitation to assure their safety. Even supermarkets are becoming involved in this development as they fight to regain lost market share to the fast food chains.[62] Containing precut vegetables, condiments, and other food items, salad bars are popular with consumers who like the convenience, nutrition, and appeal. Watching the proliferation of these food displays, health officials are concerned about maintaining adequate sanitation. In 1984 the New York State Department of Agriculture and Markets proposed salad bar rules that included the following requirements:

- installation of a plastic shield called a sneeze guard,
- maintenance of the refrigerated or iced display at 45°F with containers shallow enough to be kept at that temperature,
- replacement of perishable foods every few hours (two hours is recommended and four hours is considered the maximum),
- provision of tongs or spoons for each item.[63]

Bulk Food Sales

The sale of bulk food from bins is another form of customer self-service that poses problems similar to salad bars. With a nostalgia for the old-fashioned cracker barrel, shoppers are increasingly attracted to this retail practice. Ranging from coffee, nuts, candy, spice, and cereal to dozens of other staples, bulk food is being sold as it was a century ago. Even honey and syrup are being dispensed from spigots. All this activity might appear innocent enough, except for serious sanitation questions raised by public health officials. The Colorado Department of Health is adamantly opposed to bulk food sales, but it was overruled in a 1984 court decision.[64] Meanwhile FDA has expressed concern and is reported developing guidelines for the sale of bulk food. The main points to be addressed by the guidelines are expected to include the following.

- Containers holding bulk foods should carry labels listing ingredients, including chemicals, preservatives and artificial colors or flavors (unpackaged bakery products made on the premises are excluded).

- Pet food and other items not for human consumption should be separated from other food by a barrier or open space.

- Bulk-food containers should have close-fitting, self-closing covers. The displays should be at least 30 inches from the floor.

- Dispensing utensils such as tongs, scoops and ladles should be attached to containers by tethers of such length that the utensils cannot touch the floor. When not in use the utensils should be stored in sleeves or other protective housing attached to the containers. The utensils should be cleaned and sanitized at least daily.

- Facilities where customers can wash their hands should be available.[65]

To illustrate the difficulty in anticipating all problems associated with bulk food sales, an issue was raised in 1985 about the refilling of take-home beverage containers. This practice has received a nod of approval from FDA but with a number of conditions in regard to the type of drink dispensed, equipment on hand, and instructions for safe use.[66]

Truth-in-Menu

Truth-in-menu is a food service responsibility that is sometimes stretched. Acknowledging this obligation, the Michigan Restaurant Association set a precedent in 1977 by pledging that the food and beverage which its members serve will be just what the menu says it is. In support of this position the association formulated a policy whereby a restaurant:

- will not knowingly list any item on their menus by a false or misleading name or description;
- will not substitute another item for one listed on the menu, without fully informing the customer;
- will buy only from suppliers who give a true and accurate description of items they supply to the members;
- will be scrupulously accurate in serving true and accurate weights and quantities of the items when such measure is specified on their menus; weight specified on menus being traditionally understood to mean weight before cooking.[67]

Among the misrepresentations which this policy hoped to correct was margarine labeled as butter, "bar liquor" sold under brand names, whipped toppings implied as being whipped cream, frozen fish substituted for fresh fillets, commercial foods passed off for homemade dishes, and domestic products represented as foreign appellations.

Not satisfied with voluntary compliance, various states and localities have passed mandatory truth-in-menu ordinances. In 1979 Connecticut became the first state to enforce strict regulations on the accuracy of restaurant menus. Violation of these provisions are considered breaches of the state's Unfair Trade Practices Act.[68] Singling out hamburgers for special attention, California has decreed that imitation hamburgers, e.g., those extended with soy, may not be sold as hamburgers or any cognate of this term.[69] So far the federal government has refrained from proposing any controls over menus, but FDA pointedly has warned the food service industry to be responsive to such issues as consumers' desires for more information on salt and other additives.[70]

Institutions

Perhaps it is inappropriate to make generalizations about an industry as large and varied as food service. Throughout the industry, however, there are common concerns relating to food storage, food preparation, sanitation, and employee training. Failures in any one of these areas can have dire consequences. Nowhere is this truer than in institutions, including schools, hospitals, nursing homes, and airlines, all of which have captive clientele. The following contemporary account of food poisoning is a vivid illustration of the ramifications of gross negligence.

On the dates of March 12-14, 1984, between 120 and 130 passengers and crew on British Airways flights from London to the United States were stricken with a particularly virulent strain of *Salmonella*. Other passengers on flights to destinations all over the world reported similar symptoms, probably making this food poisoning the worst airline epidemic in history. Because of the long incubation period, the cause was not immediately determined and corrected. When the evidence finally had been gathered, the contamination was traced to aspic glaze on hors d'oeuvres prepared in the airline's kitchen at Heathrow Airport. Fortunately most of the pilots were spared the illness because of an informal standing rule to serve the flight officers separate meals. Still, the outbreak was so severe that 25 percent of the afflicted persons in the United states were hospitalized, and three months after the incident one person was reported still on the critical list.[71,72]

REFERENCES

1. *1983 Annual Report*, Safeway Stores, Inc., Oakland, California.
2. *1983 Annual Report*, McDonald's Corporation, Oak Brook, Illinois.
3. Carol Kurtis, "Bigger Slice of the Pie," *Barron's*, March 29, 1976, pp. 11, 16.
4. Harold Seneker, "The New Food Giants?" *Forbes*, April 1, 1977, pp. 53, 54.
5. *Inspecting Incoming Food Materials*, DHEW Publication No. (FDA) 76-2017, Food and Drug Administration, Washington, D.C.
6. S. G. Taylor, "Trends in Materials Management," *Chemical Engineering*, Deskbook Issue, October 30, 1978, pp. 109-114.
7. "Computer Network Solves Warehouse Inventory Accounting Problem," *Food Engineering*, December, 1978, p. 190.
8. *Voluntary Industry Sanitation Guidelines for Food Distribution Centers and Warehouses*, Association of Institutional Distributors, et al., in cooperation with the Food and Drug Administration, Washington, D.C., 1974.
9. *Ten Rules for Food Warehousemen*, FDA Poster No. 1, Food and Drug Administration, Washington, D.C., 1966.
10. P. J. Mudar, "Heuristic Programming," *Chemical Engineering Progress*, December, 1969, pp. 20-24.
11. Robert Jeffrey, "Overview of Product Abuse in the Food Distribution System from a Distribution Perspective," a paper presented at the annual meeting of The American Institute of Chemical Engineers, Chicago, Illinois, November 14, 1985.
12. Jeremy Main, "The Trouble with Managing Japanese-Style," *Fortune*, April 2, 1984, pp. 50-56.

13. A. Kramer, "Changes in Food Quality 1927-1977," *J. of Food Quality*, April, 1977, pp. 1-4.
14. Byron H. Webb and Earle O. Whittier, *Byproducts From Milk*, The Avi Publishing Co., Westport, CT, 1970, p. 148.
15. Theodore P. Labuza and Mary K. Schmidl, "Accelerated Shelf-Life Testing of Foods," *Food Technology*, September, 1985, pp. 57-64, 134.
16. "Extending the Shelf Life of Fresh Foods by Combining Controlled Atmospheres and Refrigeration," *Food Technology*, March, 1980, pp. 44-71.
17. Neil H. Mermelstein, "Hypobaric Transport and Storage of Fresh Meats and Produce Earns 1979 IFT Food Technology Industrial Achievement Award," *Food Technology*, July, 1979, pp. 32-34.
18. "Carbon Dioxide Kills Bugs in Grain Storage Bins," *Chemical Week*, June 13, 1979, p. 23.
19. Michael F. Veranth, Karl Robe, "CO_2-Enriched Atmosphere Keeps Fish Fresh More Than Twice As Long," *Food Processing*, April, 1979, pp. 76-79.
20. Larry Bell, "Gas in Trucks, Containers Helps Packages Keep Fresh Foods Fresh," *Package Engineering*, March, 1980, pp. 72-75.
21. "Carbon Dioxide Inhibits Cantaloupe Decay," *Food Engineering*, May, 1979, p. 224.
22. Ronald Richter, "Atmosphere Control Reduces Bacteria," *Dairy Field*, November, 1982, pp. 78, 79.
23. "Preserving Fresh Foods Without Preservatives," *Business Week*, September 24, 1984, pp. 144H-144L.
24. Stanley Elias, "UPC: Where's It Going?," *Food Engineering*, April, 1979, pp. 80, 81.
25. "Tomorrow's Technology Nears for Universal Case Code," *Package Engineering*, November, 1981, pp. 31-35.
26. "The Codemakers," *Forbes*, July 30, 1984, pp. 12, 13.
27. "Single-Label Weigh/Price/UPC-Symbol Printer Gives Processors/Supermarkets Better Marketing Linkup," *Food Processing*, July, 1980, pp. 49, 50.
28. "Bar Coding Is Cash-Register Music to This Association of Suppliers," *Food Engineering*, October, 1984, p. 21.
29. "Schmidt Sets 90-Day Deadline for Voluntary Clean Up of Food Railcars," *Food Chemical News*, September 16, 1974, pp. 42-44.
30. "Trucks, Railcars Have Equal Defective Rates in FDA 1977 Program," *Food Chemical News*, November 20, 1978, pp. 40-43.
31. "FDA to Press ICC for Better Quality Control of XF Boxcar Fleet," *Food Chemical News*, December 12, 1977, pp. 35, 36.
32. "ICC Study Finds Food Debris in Railroad Cars and Trucks," *Food Engineering*, September, 1978, pp. 49, 50.
33. "Nearly Two-Fifths of Boxcars Offered for Food Hauling Have Defects," *Food Chemical News*, October 8, 1979, pp. 29, 30.
34. "ICC Is Pressed to Develop New Rules for Food Uses of Rail Cars," *Food Chemical News*, August 27, 1979, pp. 16, 17.

35. "FDA Lacks Recourse Against Railroad Cars Rejected by Shippers, Cooper Says," *Food Chemical News*, January 14, 1980, p. 36.
36. Mel Seligsohn, "Railroad Car Sanitation Stamped 'Top Priority' by FDA," *Food Engineering*, July, 1978, pp. 70, 71.
37. "Better Sanitation in Food Transport Vehicles Sought in FDA 'Action Program'," *Daily Traffic World*, April 11, 1977.
38. "Railcar Sanitation Committee Changes Name, Expands Scope," *Food Chemical News*, March 27, 1978, pp. 21, 22.
39. "Transportation Guidelines List Shipper, Receiver & Carrier Roles," *Food Chemical News*, October 11, 1976, pp. 12, 13.
40. *Voluntary Transportation Guidelines*, American Feed Manufacturers Association, et al., in cooperation with the Food and Drug Administration, Washington, D.C., 1976.
41. "Rail Car Committee Supports Exclusive Use of XF Boxcars for Food," *Food Chemical News*, August 1, 1977, pp. 29, 30.
42. "A New Ploy to Eliminate Deadheading," *Chemical Business*, March 10, 1980, pp. 49, 50.
43. "Food-by-Mail: New Market for the '80s?" *Food Engineering*, January, 1981, pp. 70-72.
44. "Our 'Miniature Mailable Freezer'," Sales brochure from Omaha Steaks International, Omaha, Nebraska, 1984.
45. Roger W. Miller, "For Your Dining Pleasure — A Model Ordinance," *FDA Consumer*, February, 1980, pp. 17-19.
46. James Greene, "FDA's New Code: A Clean Sweep for Food Stores," *FDA Consumer*, October, 1982, pp. 12, 13.
47. "FMI Accepts Revision of Retail Food Store Sanitation Code," *Food Chemical News*, June 21, 1982, pp. 39, 40.
48. "Proposed Retail Food Store Model Ordinance Is Available for Comment," *Food Chemical News*, October 31, 1977, pp. 8-12.
49. "Structure of New Retail Food 'Unicode' Agreed to by Task Force," *Food Chemical News*, January 5, 1987, pp. 30, 31.
50. *The Vending of Food and Beverages*, Public Health Service, Washington, D.C., 1965.
51. *Food Service Sanitation Manual*, DHEW Publication No. (FDA) 78-2081, Food and Drug Administration, Washington, D.C., 1976.
52. *Retail Food Store Sanitation Code*, Association of Food and Drug Officials, York, PA, and Food and Drug Administration, Washington, D.C., 1982.
53. "USDA Lowers 'Safe' Pork Internal Temperature to 160°F," *Food Chemical News*, June 23, 1986, pp. 24, 25.
54. Chester G. Hall, "National Program for Food Service Sanitation," *Food Technology*, August, 1977, p. 72.
55. E. H. Marth, "Public Health and Regulatory Concerns of the Foodservice Industry," paper No. 106 presented at the 1984 Annual Meeting of the Institute of Food Technologists, Anaheim, CA.

56. "NRA Opposes FDA Certification System for Foodservice Managers," *Food Chemical News*, October 22, 1984, p. 19.
57. "First Round of Food Protection Certification Testing to Begin Next Month," *Food Chemical News*, July 15, 1985, p. 14.
58. Leon J. Marano, "The Effect of Regulations on the Food Service Industry — The Operator's Viewpoint," *Food Technology*, August, 1977, p. 73.
59. John R. Janke, "Coping with Local and Federal Regulations," *Food Technology*, August, 1977, pp. 75, 76.
60. J. R. Stubblefield, "Significance of Federal, State, and Local Regulatory Agencies on a Multi-Unit Foodservice Company," paper No. 108 presented at the 1984 Annual Meeting of the Institute of Food Technologists, Anaheim, CA.
61. C. F. Bryson, "Federal/State Regulatory Impact on Foodservice Industry," paper No. 104 presented at the 1984 Annual Meeting of the Institute of Food Technologists, Anaheim, CA.
62. "The Supermarkets Fight Back," *Dun's Review*, October, 1977, pp. 108-110.
63. Florence Fabricant, "Proliferation of the Salad Bar," *The New York Times*, May 23, 1984, pp. C1, C12.
64. "Colorado Asks Food Chains Not to Sell Bulk Foods," *Food Chemical News*, October 8, 1984, pp. 5-7.
65. Bryan Miller, "As Bulk Food Sales Grow, So Do Health Questions," *The New York Times*, January 18, 1984, pp. C1, C8.
66. "FDA Approves Use of Beverage Refilling Containers in Retail Outlets," *Food Chemical News*, September 9, 1985, pp. 18, 19.
67. "Truth in Menu Policy Adopted in Michigan," *Food Product Development*, April, 1977, p. 12.
68. "Connecticut to Enforce Truth-in-Menu Regulations by March 2," *Food Chemical News*, January 15, 1979, pp. 21, 22.
69. *California Restaurant Act*, Department of Health, State of California, 1975.
70. "Miller Tells Food Service Industry to Respond to Public's Concerns," *Food Chemical News*, August 20, 1979, pp. 28, 29.
71. Carole A. Shifrin, "Health Service Traces Food Poisoning," *Aviation Week & Space Technology*, April 2, 1984, pp. 29, 30.
72. Lawrence K. Altman, "Publicity about Airline Incident Leads to Crucial Diagnosis," *The New York Times*, June 26, 1984, p. C3.

CHAPTER THIRTEEN
PRODUCT RECALL

"Recall is quality control when it's too late. It is like taking out an advertisement on national TV to make sure that everybody knows you 'goofed'." Those words were spoken by Alexander M. Schmidt, Commissioner, Food and Drug Administration (FDA), at a product recall seminar sponsored by The Conference Board in 1974. Notwithstanding a ring of truth in these observations, Dr. Schmidt went on to acknowledge that as long as man and his machines remain fallible, product recall will be "a legitimate and essential part" of any effective quality assurance program. "Most [persons] in industry accept recall as a necessary technique to be used openly and honestly with consumer and with government participation."[1]

Preparation for a product recall must begin long before a company is faced with this unpleasant task. A procedure must be approved outlining the recall organization that will handle the emergency (cf. Chapter One). The individuals who are in charge of each function should be identified and their responsibilities ought to be described in detail. A directory of these persons, including office and home telephone numbers, needs to be included in any product recall manual. Other preparatory work is required besides the development of a recall procedure. Meaningful coding of products must be established, and a record-keeping system must be adopted so that individual lots of material can be traced from the manufacturing plant to the end user. Without this set of controls in place, a company, in a worst case scenario, may be forced to recall everything!

CODE DATE

Product coding is accomplished by using code dates to identify units of production. The code date, sometimes called a key date, is a number which contains three important pieces of information: the date on which the product was manufactured, its batch or lot number, and the identity of the facility which produced it. In conjunction with the product number, the code date incorporates all the vital information about a particular food item.[2,3] This code can take several forms, but preferably it should conform to the convention of the food industry. (Code dates used for food products correspond to serial numbers for durable goods.) The object of an effective code date is to contain all the pertinent data, be as brief as possible, and at the same time be easy to decode.

Alphanumerical Format

The Grocery Manufacturers of America, Inc. (GMA) suggests that the code date be composed of an alphanumerical format containing eight digits or letters.[4] The first three figures indicate the Julian date or the day of the year when the material was produced. The fourth place is reserved for the plant identity, which is shown by a letter. In the fifth place is a number to designate the year. The last three digits indicate the lot or batch number. One of them may be a letter, for example, to show the packaging line. Also, the shift and hour may be substituted for the lot number. To illustrate the GMA code containing a lot number, a product from lot number 17, produced on June 17, 1987, at a manufacturing facility located in Chicago could be identified as follows:

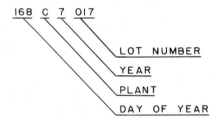

Many manufacturers use variations of the GMA code derived by transposing some of the specified information or by shortening the lot number. Instead of these various forms, an industry-wide standardized code would be easier to recognize and adapt to automatic data processing equipment.

GMP Regulations

As specified above, the code date provides positive identification of lots or batches of any given food product. This requirement was spelled out in the former Current Good Manufacturing Practice (GMP) regulations thus:

> Meaningful coding of products sold or otherwise distributed from a manufacturing, processing, packing, or repacking activity should be utilized to enable positive lot identification to facilitate, where necessary, the segregation of specific food lots that may have become contaminated or otherwise unfit for their intended use. Records should be retained for a period of time that exceeds the shelf life of the product, except that they need not be retained more than 2 years. (21CFR110.80[i])

When the GMP regulations were reissued on June 19, 1986, coding and record-keeping requirements were glaringly absent. FDA explained that it was bowing to objections raised about costs, but that it nevertheless "encourages firms to code their products and to maintain appropriate records." In a survey FDA found that 95 percent of large manufacturers sampled and 93 percent of the small ones coded their products. The agency further stated that those manufacturers which choose to ignore coding do so at their own risk. FDA concluded:

> That an industry confronted with little likelihood of recalls of products subject to the proposed rule could decide that removal of all offending products from the market in the presence of a recall would protect the public health and would be more cost effective than maintaining records and coding products. On the other hand, an industry confronted with a high frequency of recalls or with the apparent potential for infrequent but serious contamination of a limited quantity of product, could decide that coding and record-keeping are essential to accomplish a recall.[5]

Application and Entry of Code Dates

For effective product coding, every package or container should be marked with the appropriate code date. It should be located where it can easily be found, e.g., at one of the lower corners of the principal display panel, and it should be legible so as not to create confusion. For rapid identification, the code date should be preceded by the words, "code date," "lot number," or other suitable prefix. In addition, some manufacturers choose to number sequentially every container or carton within a lot in order to maintain even tighter control over distribution. The only exception to marking each container with a code date is in the case of a private labeled product, but even here some designation such as the customer's coding should be used.

The code date should be entered on all quality control records, production reports, and shipping forms. Unfortunately, once the product leaves the control of the manufacturer, code dates are not always recorded on shipping instructions. Instead of using code dates, public and customer warehouses often rely on such data as the month and year when a product is received in order to control stock rotation.[6] By contrast, the distribution of durable goods may be traced with greater ease by enclosing postage-paid registration or warranty cards that the customer is asked to return.[7] Perhaps, standardization of the code date format would permit improved control over the distribution of food products at a minimal additional cost.

Bulk Shipments

Bulk shipments present a special problem for coding. Rather than using conventional code dates, these shipments may be identified by date of shipment and the number of the tank car, tank truck, hopper car, or truck.[8] Furthermore, food manufacturers are advised to segregate product in bulk storage facilities by a limited number of lots or batches whenever possible. If this procedure is not possible, manufacturers should not commingle production lots or bulk receipts until a complete analysis has been obtained. As soon as there is complete assurance that a lot is in compliance, it may be transferred from a holding tank or bin to bulk storage. Because the quantities of material involved in bulk shipments are often huge, sometimes as much as a boatload, these movements require extra care.

Product Recall Card

Once a coding procedure has been developed, this information can be used to establish the complete records needed in the event of a product recall. A card as shown in Figure 13.1 should be prepared for every lot of food product manufactured or packaged. Each card indicates the lot size and, when completed, shows where every container in the lot has been sent. Traditionally a modest size business would have filed these records on standard index cards. Now this information may be stored by a personal computer which is capable of searching for the necessary data. The important consideration is that once a code date is specified, all the consignees of that lot can be retrieved along with the quantities of the material shipped and the dates when shipments were made. Reverse access can also prove to be useful so that for any given consignee specified, all the code dates of product shipped to the account can be determined.

RECALL POLICY

To the best of anyone's knowledge, product recall is a management device that seems to have had a spontaneous origin. Although little is known about its early history, product recall has evolved into a stylized procedure that is one of the most effective corrective measures wielded by industry. Euphemistically called "distribution in reverse," product recalls have become so commonplace that these actions are now received by the public with a degree of ennui. Even so, this procedure has been incorporated into the regulations of just about every consumer product sold.[9] Two landmark bills, the National Traffic and Motor Vehicle Safety Act of 1966 and the Consumer Product Safety Act of 1972, have contributed to the substantial increase in the number of mandatory recalls which are conducted by American business.[10]

FDA Voluntary Guidelines

Until the passage of the Infant Formula Act in 1980, the Federal Food, Drug, and Cosmetic Act (FD&C Act) was moot on the subject of product recalls.[11,12] This amendment, however, only covers a specific class of food products so that foods in general escaped any legal requirement pertaining to recalls. Notwithstand-

PRODUCT RECALL CARD

Product _____ MCP _____   Product No. ___ 32069 ___

Code Date ___ 122C4724 ___   Lot Size ___ 200 50-Lb. bags

Name of Consignee	Quantity Shipped	Date of Shipment	Inventory Balance
Minn. Flour	100	5-18	100
World Bakery	75	5-21	25
Gen. Cookie	20	5-24	5
First Food	5	5-25	0

FIGURE 13.1. RECALL CARD

ing this limitation, FDA went ahead in 1978 and finalized guidelines for conducting product recalls related to all foods. Codified as a voluntary procedure, this regulation allows that a product recall may be initiated by a food manufacturer or it may be undertaken by a manufacturer at the request of FDA. A policy statement asserts that ''Recall is generally more appropriate and affords better protection for consumers than seizure, when many lots of product have been widely distributed.'' (21CFR7.40)

FDA was spurred to issue guidelines for recalls because of the indiscriminate growth of such actions and the lack of any clear standards for their execution. The agency observed a wide discrepancy in the effectiveness of recalls conducted by food manufacturers, who often were ill-equipped to handle the necessary intricacies.[13] At the same time FDA appreciated the fact that voluntary recalls, when properly conducted, had clear-cut advantages, many of which redounded to the benefit of consumers. The agency concluded that "an FDA-initiated recall is the action of choice (if [a] firm has not already initiated a recall)" in its enforcement program.[14]

The voluntary nature of product recalls demands a cooperative attitude from industry. A blend of altruism and enlightened self-interest is the principal motivation for such cooperation. This need was recognized in the preamble to FDA's proposal to issue voluntary guidelines:

> Most firms honor FDA requests to recall violative foods. . . . Such action, whether the result of a specific request by FDA or initiated by the firm, may be prompted by several motives: The firm's sense of responsibility, its obligation to prevent harm to the public health and welfare, its desire to avoid an FDA-initiated legal action, or its desire to minimize civil liability. Whatever the motivation may be, the vast majority of firms manufacturing or distributing FDA-regulated products carry out their responsibility to remove or correct distributed products that are defective.[15]

USDA Policy

The U.S. Department of Agriculture (USDA) has followed FDA's lead in promulgating product recall guidelines. Back in 1975 USDA and FDA signed a Memorandum of Understanding (MOU) covering secondary recalls that USDA would request of a manufacturer when FDA initiates recalls for any ingredients used in meat and poultry products.[16] This MOU has been followed by a directive (8080.1) from USDA's Food Safety and Inspection Service for product recalls. This directive closely adheres to FDA's guidelines except for the criteria concerning public notification. Because recall classifications and other terms are similarly defined by FDA and USDA, there should be a minimum of confusion posed by two sets of guidelines.[17]

RECALL CLASSIFICATION

Chances are that the first tip-off of the need for a recall will come from a customer product complaint (cf. Chapter Ten). The responsibility for spotting a problem rests with the Director of Quality Assurance. He will coordinate an investigation of the problem in much the same manner as for any complaint but at a greatly accelerated pace. *Speed is essential in a product recall to minimize the hazard.* As he proceeds in his investigation he should maintain detailed notes or a diary of the events as they unfold on an hour-by-hour basis.[18] If he is suspicious that a product recall may be necessary, his first action is to immediately slap a freeze or embargo on the shipment of all suspected product. Every company-operated facility and all public warehouses should be notified to set aside the product in question and tag it, "Hold — do not ship."

Determination of the Need for a Recall

The investigation of the need for an impending recall must answer two questions fast: What is the gravity of the infraction? and How widespread is the occurrence? The investigation must address all safety issues, and it should identify the code dates which may be in violation. To visualize this process, a graph may be prepared, as illustrated in Figure 13.2, of the defect versus the lot numbers. The extent to which the product deviates from specification will give some idea of the potential danger. By determining the base line just prior to and immediately after the upset, the contaminated lots can be bracketed.

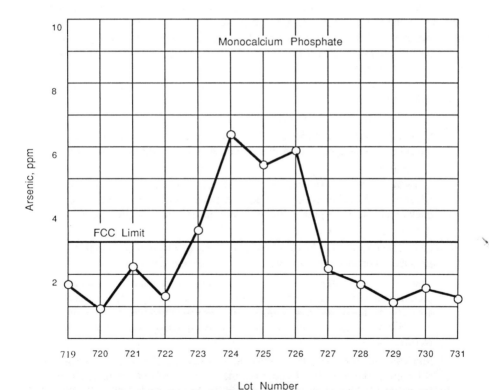

FIGURE 13.2.
BRACKETING ADULTERATED LOTS FOR PRODUCT RECALL

With the facts in hand, the Director of Quality Assurance, now acting in the capacity of the Recall Coordinator, must proceed without delay to take the necessary corrective measures. If a market withdrawal or stock recovery, as defined below, is called for, he may be empowered to initiate these actions. On the other hand, if the situation is more severe and requires a product recall, he should prepare recommendations for the general manager or chief executive officer (CEO). These proposals will include the recall classification for the recommended action. The final decision will be up to the CEO.

Definitions of Recall Classifications

A company may decide to recall product for one of several reasons. Product may be non-conforming because it is (a) "adulterated," (b) "misbranded" (both terms being defined by the FD&C Act), (c) otherwise not in compliance with an applicable government regulation, or (d) not in accord with an established specification and/or standard, whether adopted by the company, a customer, or other body. These possibilities are covered by numerical classifications which have been defined and promulgated by FDA. While the term recall is often applied loosely to all actions in which product is taken back, the regulations reserve the use of the word recall for the three most critical actions. On the basis of the health hazard evaluation, one of the following classes will be assigned to an action to recover product in order to convey in lay terms the relative severity of the problem.

> **Class I** [Recall] is a situation in which there is a reasonable probability that the use of, or exposure to, a violative product will cause serious adverse health consequences or death.
>
> **Class II** [Recall] is a situation in which use of, or exposure to, a violative product may cause temporary or medically reversible adverse health consequences or where the probability of serious adverse health consequences is remote.
>
> **Class III** [Recall] is a situation in which use of, or exposure to, a violative product is not likely to cause adverse health consequences.
>
> **Market Withdrawal** means a firm's removal or correction of a distributed product which involves a minor violation that would not be subject to legal action by the Food and Drug Administration or which involves no violation, e.g., normal stock rotation practices, routine equipment adjustments and repairs, etc.
>
> **Stock Recovery** means a firm's removal or correction of a product that has not been marketed or that has not left the direct control of the firm, i.e., the product is located on premises owned by, or under the control of, the firm and no portion of the lot has been released for sale or use. (21CFR7.3)

Interpretation of Definitions

Although the definitions for the different levels of recall appear to be direct and easy to understand, they are not always simple to apply in real situations. Take the case of a brewery which decided to pull back certain cases of its beer because of the development of an off-taste. It maintained that the removal was like any routine withdrawal to assure stock rotation. It therefore was miffed at the opprobrium it received when FDA listed the action as a Class III recall. The agency asserted that this classification was correct because the product was indeed defective.[19]

In this particular case involving off-taste beer, FDA's position could probably be justified by a strict interpretation of the FD&C Act, which defines adulterated food as including any product "otherwise unfit for food." (Section 401 [a][3]) Still, the question is left unanswered as to what constitutes a "minor violation" in the definition of a market withdrawal. FDA has been quoted as saying,

> The agency will be guided by the same considerations and principles of reasonableness that are used in deciding whether to initiate a court action; however, it is not practical to specify in these regulations in a quantifiable sense when a violation may be minor.[20]

The problem of measuring the seriousness of a "minor violation" is no different from determining the meaning of "adequate ventilation" or "excessively dusty roads" in the Good Manufacturing Practice regulations. (21CFR110.20) The final determination must be made in the courts on a case-by-case basis.

A less controversial market withdrawal was conducted in 1984 by Kellogg Company, which called back boxes of its popular brand of cereal, Corn Flakes. These boxes carried the picture of Vanessa Williams, twenty-one year old beauty queen and reigning Miss America. Kellogg was worried about the sensitivities of the younger set, which devours the greatest quantity of breakfast cereal. The concern arose when Ms. Williams was caught in some "compromising positions" by a photographer for *Penthouse* magazine and was forced to abdicate her title. The stir was too risque for the stout souls from Battle Creek, Michigan.[21,22] But regardless how important this incident was to the internal affairs of Kellogg, it was of no interest to FDA.

RECALL STRATEGY

While a recall classification does not by itself dictate the recall strategy, i.e., course of action, it is an important consideration. Other factors besides the health hazard evaluation, however, need to be taken into account. They are (a) the ease in identifying the product, (b) degree to which the product's deficiency is apparent to the user, (c) extent to which the product has been consumed, and (d) availability of necessary replacements. (21CFR7.42) The variety of circumstances surrounding a recall requires a flexible approach in the development of recall plans. For this reason a unique recall strategy should be developed for each recall.[23]

Administrative and financial responsibility for implementing a recall strategy rests completely on the shoulders of the company which manufactured the product. This entity may seek advice and assistance from FDA, but it alone is accountable.[24] Should the manufacturer show reluctance to proceed with the necessary action, FDA is empowered under Section 705 of the FD&C Act to publicize widely any "imminent danger to health, or gross deception of the consumer."[25] The manufacturer must further reckon with the terms of his product liability insurance. These policies invariably demand that sister products of a defective item be recovered so as to limit the exposure to damages.[26]

Depth of Recall

The first step in formulating a recall strategy is to agree on the depth of the recall. Three basic options are available:

1. Consumer or user level, which may vary with product, including any intermediate wholesale or retail level; or
2. Retail level, including any intermediate wholesale level; or
3. Wholesale level. (21CFR7.42)

The depth of the recall will depend on the degree of the hazard and the extent of the distribution. As guidelines the following examples have been given:

Class I Recalls . . . shall be made to the consumer level;
Class II Recalls . . . shall be made to the retail or dispensing level;
Class III Recalls . . . shall be made to the wholesale level.[27]

To achieve any of the above objectives, distributors, wholesalers, or retailers may be required to initiate sub-recalls. These actions are defined as recalls made by consignees of the firm which originated a product recall.

Communication Plan

The second step in determining a recall strategy is to devise an effective communication plan. In the case of a firm-initiated recall, the manufacturer is obligated to notify FDA immediately, notwithstanding the diffident language in the regulations, which only requests such a response. FDA involvement is not only in the public interest, but a coordinated effort stands to avoid unnecessary snafus. In the event that a firm initiates a market withdrawal, this information should be conveyed to FDA if the cause for the action is not completely obvious or self-evident.

Telecommunication is the fastest and most attention-getting means of reaching consignees. Because the marketing/sales force is most familiar with customer accounts, its efforts are needed to make these contacts and to explain the reason for the recall.[28] Telephone calls and telegrams are restricted to recalls with a manageable number of consignees. Another limitation is that telephone conversations can be misinterpreted and important points may be overlooked. Therefore the information relayed in all phone calls should be confirmed as soon as possible in writing.

Letters to customers announcing a recall should be brief and to the point. They should contain the following material:

1. A complete description of the product to be recalled, including the quantity of material sent to each consignee, the dates when it was shipped, and the lot numbers;

2. The fact that FDA has requested the recall or has been notified and is involved;

3. A disclosure of the product's defect, the risks associated with its continued use, and the classification of the recall;

4. Actions to be taken by the consignee, including cessation of further use and distribution of the product, initiation of the sub-recall if necessary, and disposition of unused material in a proper way;

5. An offer by the manufacturer to assume all costs incurred as a direct result of the recall; and

6. A self-addressed, postage-paid card for reporting to the manufacturer the amount of the product consumed and the remaining stock in inventory.

A model recall letter may be drafted in advance for review by legal counsel to make sure that it conforms to the regulations and that it avoids unnecessary admissions of guilt.[29,30] By having a form letter in its file, a manufacturer can save valuable time when an emergency strikes. Sending recall letters via registered or certified mail requesting return receipts is the best assurance that they will be received by the proper individuals and that they will be heeded. The regulations, however, specify only that the letters be sent by first class mail. To catch the addressee's attention, each recall letter should be boldly marked, preferably in red ink, on the envelope and inside with the notice, "Urgent food recall."

Communication with consumers whose identity is unknown is much more involved. A virtual dragnet must be undertaken to find the defective product. The expense in terms of manpower and resources is the reason why such a procedure is restricted to recalls with high priority. At the retail level, a search must be made of every food store in a metropolitan district or larger area in order to remove all defective product from the shelves. If there is a necessity to warn households, publicity must be disseminated in the press, on radio and television. In extreme instances FDA has taken to the streets in sound trucks to warn of the dangers. A general mobilization can be quite effective at such critical times.

EFFECTIVENESS CHECKS

Through effectiveness checks, the manufacturer which conducts a recall is expected to verify that consignees have received notices and are taking the appropriate action as requested in the notices. These follow-ups may be carried out through telephone calls, letters, or personal visits. While a personal canvass has been found to be very effective, its cost precludes more extensive use of this method.[31] The self-addressed cards enclosed with the original notices provide a means for the consignees to indicate their responses. Ultimately, the best check is a count of the product that is returned.

Level of Effectiveness Checks

The level of effectiveness checks is the percentage of contacts to be made out of the total number of consignees notified. The following levels are defined for easy reference:

> **Level A** — 100 percent of the total number of consignees to be contacted;
> **Level B** — Some percentage of the total number of consignees to be contacted, which percentage is to be determined on a case-by-case basis, but is greater than 10 percent and less than 100 percent of the total number of consignees;
> **Level C** — 10 percent of the total number of consignees to be contacted;
> **Level D** — 2 percent of the total number of consignees to be contacted; or
> **Level E** — No effectiveness checks. (21CFR7.42)

An appropriate level will be determined for each recall strategy and will reflect the seriousness of the hazard. As a rule, Class I recalls will require Level A effectiveness checks, and Class II recalls will command Level B checks.[32]

Recall Status Reports

The information which is received from the effectiveness checks should be forwarded to FDA in periodic Recall Status Reports. The frequency of these reports will be determined by circumstances, but as a guide an interval between two and four weeks has been suggested. The reports should incorporate the following specified data:

1. Number of consignees notified of the recall, and date and method of notification.
2. Number of consignees responding to the recall and communication and quantity of products on hand at the time it was received.
3. Number of consignees that did not respond (if needed, the identity of nonresponding consignees may be requested by the Food and Drug Administration).
4. Number of products returned or corrected by each consignee contacted and the quantity of products accounted for.

5. Number and results of effectiveness checks that were made.
6. Estimated time frames for completion of the recall. (21CFR7.53)

Much of this information can be summarized for quick review in a presentation as shown in Table 13.1. From the totals obtained, a simple calculation can be made of the percentage of shipped product that is returned. In this particular example, the job of recovering product was made easier by the fact that a minimum number of lots had been shipped to each consignee. This procedure, namely, including as few lots as possible in each shipment, also facilitates routine quality control testing by the customer and is therefore usually required by him.

TABLE 13.1
EFFECTIVENESS OF PRODUCT RECALL

Name of Consignee**	Ship Date	Total Bags	Bags Shipped*				Bags Returned*			
			723	724	725	726	723	724	725	726
A-B Whs.	5-23	200			200				60	
First Food	5-25	100		5		95		0		68
Gen. Cookie	5-24	20		20				20		
Minn. Flour	5-18	300	200	100			100	100		
Southern Whs.	5-29	105				105				82
World Bakery	5-21	75		75				59		
TOTAL		800	200	200	200	200	100	179	60	150

*The number of bags shipped and returned are shown by lot numbers. The lot size is 200 50-Lb. bags.
**Fictitious names have been used in this and other examples.

FDA Audit Checks

In most cases an independent appraisal of the effectiveness of a recall will be made by FDA. The agency will employ "audit checks" as distinguished from effectiveness checks to determine the level of compliance. In practice, there is little difference between the two types of checks except that audits are conducted by government inspectors.[33] FDA will officially terminate a product recall when it determines that all reasonable efforts to remove or correct product have been made in accordance with the adopted strategy. (21CFR7.55) The manufacturer, however, should not relax its efforts until the causative agent has been identified and corrected.

A published record of all Class I, II, and III recalls is issued by FDA on a weekly basis in the *FDA Enforcement Report*.[34] Aside from fulfilling a need of public policy, the report is a source of valuable information to the food industry, which should keep itself apprised of enforcement actions. These records also are the basis for interesting statistics showing the level of food and drug recalls and their trend. During the 1984 fiscal year ending October 1, there were 1,408 product recalls recorded compared with the 915 in the previous year. Over a four year period ending October 1, 1984, the number of recalls was 3,690 as against 4,158 in the prior four years. This decline was slight compared to the reduction in other actions — criminal cases, seizures, and injunctions — which fell from 2,231 to 1,114 cases over the same time span. The growing reliance on product recalls recognizes the greater effectiveness of this procedure.[35]

RECONDITIONING AND DISPOSAL

The term retroaction has been suggested as a more fitting description of what goes on under the guise of product recall.[36] The reason for this view is that product recall may involve any one of several possible actions besides returning the product. Field corrections, such as "repairs, modifications, adjustments, relabeling, inspection, or destruction of products," may be carried out without their actual physical removal or return to the manufacturer.[37] Whatever method is chosen, FDA urges that the disposition be handled expeditiously. "The longer a product is held between the date of recall and time of final disposition, the greater the chance of its accidental misuse."[38]

Isolation of Defective Product

As soon as defective product is identified, it should be isolated from acceptable food products. The recalled product should be relabeled or tagged with bright orange or red stickers and marked as "Rejected" or "Not for food use." If the quantities are small they may be picked up by personnel making effectiveness checks. Otherwise the product should be sent back to the manufacturer or assembled in a staging warehouse which preferably is one not used for normal product distribution. Generally, recalled product should not be left in the hands of the customer for him to discard. The risk is too great that the defective product will be diverted to some unauthorized use. Depending on circumstances, regulatory agents may want to witness the actual destruction or disposal of the recalled product.[39]

Reconditioning

In so far as possible, manufacturers may avail themselves of the opportunity to recondition distressed product. FDA has enunciated the following policy for reconditioning food that is adulterated by reason of being prepared, packed, or held under insanitary conditions:

> FDA will accept reconditioning proposals for foods adulterated due to storage under 402(a)(4) [of the FD&C Act] conditions when such proposals contain

provisions for removing the 402(a)(4) conditions: e.g., brushing or vacuum cleaning insect filth or rodent excreta pellets from the outside of a bag, stripping urine or bird excreta stained outer layers off of a multi-ply bag, or rebagging *and* either treating the storage area to eliminate rodent or insect infestation or moving the food to a sanitary facility.[40]

When food cannot be reconditioned for human consumption, it may be reclassified as fit for animal feed. Food that is contaminated by filth must first undergo a heat treatment to destroy pathogenic organisms. Other contaminated food that contains excessive levels of pesticides or toxicants may be acceptable for feed use. While FDA policy does not sanction blending to achieve permissible tolerance or action levels, it does allow "different standards for foods intended for human use vs. food intended for animal use."[41]

Model Food Salvage Code

Regardless of the source, much sub-standard food ends up in salvage operations which save whatever product can be reclaimed. A 1980 review estimated that there were 1,000 food salvage businesses in the United States at the wholesale and retail levels. Some of them are nonce operations which are set up for a limited time to recover goods from a single disaster, such as a fire or flood. For years, salvage firms operated "in between the laws."[42]

To remedy this disgrace, the *Model Food Salvage Code* was drafted in 1984 through the combined efforts of FDA and the Association of Food and Drug Officials. This carefully considered document has several distinguishing features:

• Salvaged food shall be labeled to indicate that the merchandise has been salvaged. Replacement labels must disclose the name and address of the salvaging outfit as well as the date of reconditioning.

• Metal cans shall be "essentially free from rust (pitting) and dents (especially at rim, end double seams and/or side seams). Leakers, springers, flippers, and swells shall be deemed unfit for sale or distribution."

• Cans of food salvaged from floods, sewer backups, or other mishaps should be cleaned and sanitized.

• Potentially hazardous foods that require freezing or refrigeration must be discarded after standing for over 4 hours at a temperature above 45°F (7°C).

• Food contaminated and/or adulterated with pesticides or other chemicals are non-salvageable.

• Good manufacturing practice regulations are specified for all salvage operations including equipment, buildings, plumbing, garbage and refuse, insect and rodent control, employee hygiene, housekeeping, lighting, and ventilation.

• Salvage operations must be licensed by local or state authorities. Permits may be suspended or revoked for good cause.[43]

Disposal

Products which cannot be reconditioned, reprocessed or diverted to other legitimate uses must be destroyed or discarded. Thanks to the passage of the 1976 Resource Conservation and Recovery Act (RCRA), the disposal of solid and other waste, once a straightforward procedure, is now hobbled with red tape. The law is designed to prevent the indiscriminate dumping of hazardous materials, but it is so comprehensive that it has an impact on almost every waste. A "cradle-to-grave" manifest system has been devised to check clandestine dumping by gypsy haulers. Because the procedures that apply for hazardous material such as mineral acids are different from the requirements for relatively harmless substances, food manufacturers should segregate waste streams by specified classifications. This precaution will significantly reduce the costs of disposal.[44,45]

WRAP-UP

The past record of product recalls would suggest that industry spends too much time troubleshooting and does not devote enough attention to product planning. This state of affairs was noted in 1974 by U.S. Senator Frank E. Moss, Chairman of the Consumer Subcommittee of the Commerce Committee. He went on to predict that:

> In short, for the reasons stated — economic costs, damage to trade name, expenses incurred in product liability litigation . . . — the number of products recalled in the years to come will begin to decrease. Manufacturers will begin to practice preventive medicine and the result will not only be more satisfactory to the public, but also to the economic well-being of the manufacturer and the nation.[46]

Regrettably this message has been slow to reach its mark. A sense of urgency is missing in the food industry, which needs to dedicate its best talent and effort to quality assurance.

REFERENCES

1. Alexander M. Schmidt, "The Challenge of Product Recall," *Managing Product Recalls*, The Conference Board, New York, 1974.
2. "Moss Reintroduces Food Surveillance Controls for FDA," *Food Chemical News*, February 17, 1975, pp. 43, 44.
3. "Coding," *Federal Register*, June 8, 1979, p. 33248.
4. *Guidelines for Product Recall*, 1st ed., Grocery Manufacturers of America, Inc., Washington, D.C., 1974.

5. "FDA Issues Revised Food Good Manufacturing Practices As Final Rule," *Food Chemical News*, June 23, 1986, pp. 30-33.
6. David E. James, "Distribution Channel Considerations," *Managing Product Recalls*, The Conference Board, New York, 1974.
7. George Fisk and Rajan Chandran, "How to Trace and Recall Products," *Harvard Business Review*, November-December, 1975, pp. 90-96.
8. Morton E. Bader, "Quality Assurance and Quality Control Part 1," *Chemical Engineering*, February 11, 1980, pp. 87-92.
9. Robert Levy with Mark Levenson, "A Record Year for Recalls," *Dun's Review*, January, 1979, pp. 28-34.
10. "Managing the Product Recall," *Business Week*, January 26, 1974, pp. 46-48.
11. Eugene I. Lambert, "Legal Considerations," *Food Technology*, October, 1972, pp. 24, 29.
12. Eugene I. Lambert, "Reviewing the Quasi-Statutory Remedies," *Food Product Development*, June, 1976, pp. 32-36.
13. Sam D. Fine, "Definition and Exposition," *Food Technology*, October, 1972, pp. 22-24.
14. *Compliance Policy Guides*, Guide 7153.07, Food and Drug Administration, Washington, D.C., 1976.
15. "Recall Policy and Procedures," *Federal Register*, June 30, 1976, p. 26924.
16. "APHIS, FDA Sign Recall Agreement," *Food Technology*, August, 1975, p. 104.
17. "USDA Limits Routine Disclosure of Recalls to Class I Actions," *Food Chemical News*, September 12, 1983, pp. 19-21.
18. Channing H. Lushbough, "When the Prime QC System Fails — Recalls, Market Withdrawals and Stock Recoveries," *Food Product Development*, September, 1980, pp. 54, 55.
19. "Schlitz Questions FDA Publicity on Beer Recall," *Food Chemical News*, September 22, 1975, pp. 17, 18.
20. "FDA Says Seizure Is Appropriate Response to Ineffective Recall," *Food Chemical News*, June 26, 1978, pp. 16-20.
21. Deborah Caulfield, "Vanessa Williams May Have Lost Her Crown but Certainly Not Her Show Business Career," *Greenwich Time*, July 30, 1984, p. B3.
22. Jay Cocks, "There She Goes, Miss America," *Time*, August 6, 1984, p. 61.
23. Joseph P. Hile, "Recalls — A Continuing Concern for All of Us," a paper given at the annual convention of the National Canners Association, New Orleans, 1977.
24. William Grigg, "Recalls, The Media and Motherhood," *FDA Consumer*, November, 1982, p. 9.
25. Jerome Bressler, "What FDA Expects during a Recall," *Food Product Development*, April, 1980, pp. 64, 65.

26. James W. Peters, "Facing Product Liability and Withdrawal Risks," *Food Product Development*, May, 1975, pp. 78, 80.
27. *Regulatory Procedures Manual*, 5-00-40, Food and Drug Administration, Washington, D.C., 1973.
28. James D. Snyder, "How to Survive a Product Recall," *Sales Management*, October 14, 1974, pp. 23-27.
29. *Product Recall: What You Should Know and Do*, The Travelers Insurance Companies, Hartford, CT, 1975.
30. Michel A. Coccia, "Product Recall Notice in the Court Room," *Chemtech*, February, 1975, pp. 100-102.
31. "FDA Finds Mail, Phone Most Effective Checks on Recalls," *Food Chemical News*, April 10, 1978, p. 5.
32. *Regulatory Procedures Manual*, op. cit.
33. "Recall 'Effectiveness Checks,' 'Audit Checks' Differentiated," *Food Chemical News*, January 14, 1980, p. 24.
34. *FDA Enforcement Report*, Food and Drug Administration, Rockville, Maryland.
35. "U.S. Action on Foods, Drugs Dips," *The New York Times*, October 24, 1984, p. C14.
36. James W. Hulse, "Retroaction Vs Recall," *Food Product Development*, June, 1976, p. 36.
37. "FDA Proposes Recall Regulations," *Food Chemical News*, July 5, 1976, pp. 40-46.
38. Communication from Food and Drug Administration, December 7, 1979.
39. Charles Livingston, "Removal and Disposition," *Food Technology*, October, 1972, pp. 36-38.
40. *Compliance Policy Guides*, Guide 7153.15, Food and Drug Administration, Washington, D.C.
41. "FDA Provides for Requests for Diversion of Adulterated Foods," *Food Chemical News*, January 25, 1982, pp. 12-14.
42. Roger W. Miller, "Setting Forth the Right Way to Sell Salvaged Food," *FDA Consumer*, April, 1980, pp. 6, 9.
43. *Model Food Salvage Code*, Association of Food and Drug Officials, York, PA, and the Food and Drug Administration, Washington, D.C., 1984.
44. John W. Lynch, "The New Hazardous-Waste Regulations — Part I," *Chemical Engineering*, July 28, 1980, pp. 55-59.
45. "At the 11th Hour, Congress Cracks Down on Hazardous Wastes," *Business Week*, October 22, 1984, p. 112D.
46. Frank E. Moss, "The Manufacturer's Role in Product Safety," *Managing Product Recalls*, The Conference Board, New York, 1974.

GLOSSARY OF ABREVIATIONS

AAFCO	Association of American Feed Control Officials
AAP	American Academy of Pediatrics
ACGIH	American Conference of Governmental Industrial Hygienists
ADI	Acceptable Daily Intake
AFDO	Association of Food and Drug Officials
AIB	American Institute of Baking
AMS	Agricultural Marketing Service
ANSI	American National Standards Institute
AOAC	Association of Official Analytical Chemists
AQL	Acceptable Quality Level
ASME	American Society of Mechanical Engineers
ASQC	American Society of Quality Control
ASTM	American Society for Testing and Materials
BAM	*Bacteriological Analytical Manual*
BATF	Bureau of Alcohol, Tobacco and Firearms
BHA	butylated hydroxyanisole
BHT	butylated hydroxytoluene
BISSC	Baking Industry Sanitation Standards Committee
CA	controlled atmosphere
CAS	Chemical Abstracts Service
CFR	*Code of Federal Regulations*
CHEMTREC	Chemical Transportation Emergency Center
CIP	clean-in-place
COA	Certificate of Analysis
COP	cleaned-out-of-place
CPM	Critical Path Method
CPSC	Consumer Product Safety Commission
CQAP	Cooperative Quality Assurance Program
DAL	Defect Action Level
DES	diethylstilbestrol
DISPAC	House Science and Technology's Subcommittee on Domestic and International Scientific Planning, Analysis, and Cooperation

DOC	Department of Commerce
DOD	Department of Defense
DOT	Department of Transportation
EDB	ethylene dibromide
EDRO	Executive Director of Regional Operations
EEC	European Economic Community
EEO	Equal Employment Opportunity
EIR	Establishment Inspection Report
EPA	Environmental Protection Agency
EVOH	ethyl-vinyl alcohol copolymer
FAAA	Federal Alcohol Administration Act
FAO	Food & Agriculture Organization
FASEB	Federated American Societies for Experimental Biology
FBI	Federal Bureau of Investigation
F.C.C.	*Food Chemicals Codex*
FDA	Food and Drug Administration
FD&C Act	Federal Food, Drug, and Cosmetic Act
FEMA	Flavor and Extract Manufacturers' Association
FIFO	First In First Out
FIFRA	Federal Insecticide, Fungicide, and Rodenticide Act
FIPS	Federal Information Processing Standards
FMI	Food Marketing Institute
FMIA	Federal Meat Inspection Act
FOI	Freedom of Information Act
FPI	Food Processors Institute
FPLA	Fair Packaging and Labeling Act
FR	*Federal Register*
FSIS	Food Safety & Inspection Service
FSQS	Food Safety & Quality Service
FTC	Federal Trade Commission
GAO	General Accounting Office
GLP's	Good Laboratory Practices
GMA	Grocery Manufacturers of America
GMP's	Good Manufacturing Practices
GRAS	Generally Recognized As Safe
GTP	Good Transportation Practice
HACCP	Hazard Analysis and Critical Control Points
HEW	Department of Health, Education and Welfare
HHS	Department of Health and Human Services
HTST	High-Temperature, Short-Time
IAEA	International Atomic Energy Agency
IATA	International Air Transportation Association
ICC	Interstate Commerce Commission

ID	identification
IFC	Infant Formula Council
IFT	Institute of Food Technologists
IMCO	Intergovernmental Maritime Consultative Organization
INQ	Index of Nutritional Quality
IPB	Industry Program's Branch
IQAAP	Industry Quality Assurance Assistance Program
IRLG	Interagency Regulatory Liaison Group
IRPTC	International Register of Potentially Toxic Chemicals
ISSC	Interstate Shellfish Sanitation Conference
JECFA	Joint Expert Committee for Food Additives
JECFI	Joint Expert Committee on Food Irradiation
JND	just-noticed difference
K	Kosher
LA	low alcohol
LCL	lower control limit
LIFO	Last In First Out
MAV	Maximum Allowable Variations
MDR	Minimum Daily Requirements
MOU	Memorandum of Understanding
MPN	most probable number
MSDS	Material Safety Data Sheets
MSM	Mechanically Separated Meat
NAS	National Academy of Sciences
NASA	National Aeronautics and Space Administration
NBS	National Bureau of Standards
NCIMS	National Conference of Interstate Milk Shipments
NFPA	National Food Processors Association
NIFI	National Institute for the Foodservice Industry
NIOSH	National Institute of Occupational Safety and Health
NLT	not less than
NMFS	National Marine Fisheries Service of the U.S. Dept. of Commerce
NMT	not more than
NRA	National Restaurant Association
NRC	National Research Council
NSSP	National Shellfish Sanitation Program
NTP	National Toxicology Program
OM	open mouth (bag)
OMB	Office of Management and Budget
OSHA	Occupational Safety and Health Administration
OTC	over-the-counter
PBB	polybrominated biphenyls

PCB's	polychlorinated biphenyls
PDP	Principal Display Panel
PE	polyethylene
PERT	Program Evaluation and Review Technique
PET	polyethylene terephthalate
PMN	Premanufacture Notification
PMO	Pasteurized Milk Ordinance
ppb.	parts per billion
ppm.	parts per million
PUFI	Packed Under Federal Inspection
QA	Quality Assurance
QAU	Quality Assurance Unit
quats	quaternary ammonium compounds
RCRA	Resource Conservation and Recovery Act
RF	radio frequency
RODAC	replicate organism direct agar contact
RPM	Regulatory Procedures Manual
SI	Surveillance Index
SOM	Sensitivity of Method
TLV	Threshold Limit Value
TM	Trademark
TNTC	Too Numerous To Count
TQC	Total Quality Control
TRR	Trade Regulation Rules
TSCA	Toxic Substances Control Act
UCL	upper control limit
UHT	ultra-high temperature
UPC	Universal Product Code
USDA	U.S. Department of Agriculture
USDC	U.S. Department of Commerce
USDL	U.S. Department of Labor
U.S. RDA	U.S. Recommended Daily Allowances
WHO	World Health Organization
WIC	Supplemental Food Program for Women, Infants, and Children
WONF	With Other Natural Flavors
XF	Exclusively-for-Food

INDEX